Quality Management in Reverse Logistics

Yiannis Nikolaidis
Editor

Quality Management in Reverse Logistics

A Broad Look on Quality Issues
and Their Interaction with Closed-Loop
Supply Chains

 Springer

Editor
Yiannis Nikolaidis
University of Macedonia
Naoussa
Greece

ISBN 978-1-4471-4536-3   ISBN 978-1-4471-4537-0   (eBook)
DOI 10.1007/978-1-4471-4537-0
Springer London Heidelberg New York Dordrecht

Library of Congress Control Number: 2012949085

Printed on acid-free paper

Springer is part of Springer Science+Business Media (www.springer.com)

To Aristotle and Mariana

Preface

The evolution of Reverse Logistics (RL) has been remarkable lately. In the last decades, not only in academia, but also in the world of businesses, RL has started to bloom. More specifically, in the manufacturing sector the fact that companies get increasingly involved in RL should be attributed mainly to (i) environmental concerns and the pursuit of sustainability, (ii) the implementation of legislation that dictates the production and sale of environmentally friendly products, and (iii) the fact that companies become progressively aware of the profitability of the various reuse activities.

However, the RL level of organization should rise much further. For example, RL activities are vulnerable in various types of information inaccuracies, such as the uncertainty with respect to timing, quantity, and quality of product returns. So far, the research community has made significant efforts and progress to reduce the impact of these uncertainties. However, there is still a long way to go.

Focusing particularly on the quality level of returns, one can intuitively expect that the significant growth of Quality Management (QM) in "conventional" production processes as well as the impressive amount of knowledge that has been accumulated since the appearance of the first quality Gurus, could—but most importantly should—be exploited in the world of RL as well. Statistical Quality Control, QM Systems, Total Quality Management techniques, Quality tools are only some of the methods that are expected to be exploited in RL, in the near future.

In this book we deal with various quality issues in an attempt to recognize areas where the existing knowledge on QM could be implemented—perhaps appropriately modified—in RL. Moreover, we discuss more extensively than ever the role of QM in RL and examine broadly the interaction of various QM issues with the design and processes of Closed-Loop Supply Chains (CLSCs).

The introductory chapter presents the state-of-the-art regarding the relationship and interaction between the research areas of QM and RL. More specifically, a literature review identifies the most important research articles that probe the impact of quality issues on RL.

Quality uncertainty of returned products is the main source of uncertainty and, consequently, a major determinant of creating value in CLSCs. However, there is a gap in literature regarding the identification of all quality dimensions and the ways they affect CLSC processes. Therefore, the main contribution of Corbacioglu and van der Laan is the construction of a quality framework that detects the dimensions of quality in CLSCs, defines these dimensions explicitly, and links them to various CLSC processes. Hence, the developed framework enables the analysis of CLSCs through a quality point of view and the finding of new ways of value creation.

In a chapter devoted to the contribution of standardization on RL, Pirlet first explains the main characteristics of standardization in order to specify the existing standardization bodies and the alternative standardization deliverables. Then, some important challenges in RL are analyzed and the possible structures for developing new standards are presented. A framework which can improve the efficiency in the management of RL is finally developed.

The Chap. 4 by Nikolaou and Evangelinos is dedicated to the presentation of a new methodological and conceptual framework, which consists of a set of indicators. The latter permits companies and other stakeholders to measure the social responsibility quality and performance of their RL activities. The suggested framework is based on the Triple Bottom Line approach and, more specifically, on the Global Reporting Initiative guidelines. Moreover, it helps to overcome the limitations of all the existing models, by evaluating the overall social responsibility performance and not only by designing a strategic and decision support mathematical system.

Sroufe finds out that there is lack of information in literature on the role of small demanufacturers within recycling systems and on the way these entities leverage Quality Assurance (QA) standards and certifications. Consequently, a primary objective of his chapter is to enhance the understanding of how small companies overcome emerging challenges as well as take advantage of opportunities appearing in CLSCs that focus on recycling of consumer electronics and IT assets. Moreover, conducting a field study, the author sheds new light on how QA has evolved through the use of existing ISO standards and new certifications to meet the needs of RL and recycling systems.

Emerging information technologies, such as sensors and radio-frequency identification (RFID) tags, are discussed by Ondemir and Gupta in Chap. 6. They argue that recovery decisions can be made more effectively for end-of-life products equipped with embedded sensors. The information collected and provided by these devices enables the accurate identification of the end-of-life products' quality, so that disassembly, recycling, and remanufacturing operations can be carried out in a more effective way. The authors provide an integer programming model which determines the minimum cost solution to process each and every end-of-life product on hand.

An RFID integrated QM system for a RL network is presented in the next-to-last chapter by Awasthi and Chauhan. The proposed tool integrates data collection, analytical processing, quality monitoring, and recommendations generation.

They also provide a numerical case study to demonstrate the application of the proposed product quality monitoring tool.

Finally, in the last chapter Denizhan and Konuk investigate the causes of damages in a Third-party Logistics provider and propose a framework which permits the accurate classification of damages. Overall, damages that occur in forward logistics processes constitute a significant reason why products enter the RL stream. The specific system of classifying and categorizing cases of damage enables the discovery of patterns and correlations in the incidents of damage which can be used to determine appropriate prevention actions. Consequently, it helps to manage and reduce potential damages both in forward supply chains and in CLSCs.

At this point, I would like to sincerely acknowledge the valuable help of many significant researchers who reviewed the submitted chapters and contributed through their remarks to often substantial improvements. Therefore, many thanks should be addressed to Anjali Awasthi, Marisa de Brito, Meltem Denizel, Konstantinos Evangelinos, Patroklos Georgiadis, Surendra Gupta, Anand Kulkarni, Erwin van der Laan, George Nenes, Ioannis Nikolaou, Sofia Panagiotidou, Robert Sroufe, George Tagaras, Ayse Cilaci Tombus, Dimitrios Vlachos, Ton van der Wiele, Anastasios Xanthopoulos, and Christos Zikopoulos.

Yiannis Nikolaidis

Contents

1 Reverse Logistics and Quality Management Issues:
   State-of-the-Art   1
   Yiannis Nikolaidis

2 A Quality Framework in Closed Loop Supply Chains:
   Opportunities for Value Creation 21
   Umut Çorbacıoğlu and Erwin A. van der Laan

3 Standardization of the Reverse Logistics Process:
   Characteristics and Added Value 39
   Ir André Pirlet

4 A Framework for Evaluating the Social Responsibility Quality
   of Reverse Logistics 53
   Ioannis E. Nikolaou and Konstantinos I. Evangelinos

5 Quality Assurance and Consumer Electronics Recycling 73
   Robert Sroufe

6 Quality Assurance in Remanufacturing with Sensor
   Embedded Products 95
   Onder Ondemir and Surendra M. Gupta

7 An RFID Integrated Quality Management System for Reverse
   Logistics Networks 113
   Anjali Awasthi and S. S. Chauhan

8   Cases of Damage in Third-Party Logistics Businesses   131
Berrin Denizhan and K. Alper Konuk

Index   157

Chapter 1
Reverse Logistics and Quality Management Issues: State-of-the-Art

Yiannis Nikolaidis

Abstract It is generally accepted that the positive effects of quality management are several. An indicative list should undoubtedly contain the reduction of costs which are attributed to the poor quality of products or services, better relationships with suppliers and customers, faster distribution of products or services to market, reduced waste, increased added value to customers and better working conditions. More or less, these are also the benefits that companies should expect if quality management prevails in Reverse Logistics (RL) and recovery activities. Therefore, the main target of this book is to elucidate these benefits, discuss extensively for the first time the role of quality management in RL and examine broadly various quality issues and their interaction with RL. Taking the first step into this direction we exhibit the state-of-the-art regarding the interaction of various research areas on quality management and RL, by identifying the most important research articles that have recently appeared and probe the impact of quality issues on RL.

1.1 Introduction

One of the most significant production issues that businesses have focused on during the last decades is quality. That includes not only the quality of products but also the quality of services and processes. As markets become more and more competitive, quality becomes a key ingredient for business success, as customers get more and more aware of its significance and ask from every company to assure the fulfilment of their needs.

Y. Nikolaidis (✉)
Department of Technology Management, University of Macedonia,
59200 Naoussa, Greece
e-mail: nikolai@uom.gr

Y. Nikolaidis (ed.), *Quality Management in Reverse Logistics*,
DOI: 10.1007/978-1-4471-4537-0_1, © Springer-Verlag London 2013

Among the multiple definitions of quality that can be found in literature, the most comprehensive one is the following:

"Quality is meeting or exceeding the needs and expectations of customers."

This means that quality is more than a product that simply works properly. It may also include the concepts of performance, appearance, availability, timely and proper delivery, reliability, maintainability, cost-effectiveness and low price.

Manufacturing products of the required quality does not happen by chance! There has to be a production process which is properly managed, ensuring satisfactory and consistent quality. To this end, Quality Management (QM) includes all activities ensuring that products and services fit their purpose and meet the predetermined specifications. As such, it constitutes an old scientific and research area, which has been developed and implemented in various stages of forward, mainly, production processes.

It is well known that QM can be divided into two main parts: *Quality Assurance* (QA) and *Quality Control* (QC). QC is the "old-fashioned" way of quality management. It mainly aims at detecting non-conforming output, rather than preventing it and in quite a few occasions it can be a very expensive process. Hence, businesses usually focus on QA which is actually all systematic processes ensuring that a product or service meets specified requirements. More specifically, QA is about the way a product or service is produced or delivered, aiming at minimising the chances to become non-conforming. Therefore, the focus of QA is mainly on the product design, as well as on the processes and procedures involved at the production of a product or service. If the product design and production processes are tightly controlled, then quality will probably appear at a high level. Consequently, there will be less need to conduct thorough QC.

The history and evolution of QM, from the mere inspection and QC of the past to the contemporary QA, Total Quality Management (TQM) and the various modern QM techniques, such as Six Sigma, Quality Function Deployment, etc., have led to the development of theories, processes, methods and tools that are crucial to organisational development and performance improvements.

The roots of QM go back to early 1920s when Shewhart [63] developed the inspired "Control chart". Over the years, his work has evolved due to the contribution of various researchers. In the late 1940s, Americans, such as Deming, Juran, etc., developed further the concept of QM in Japan (e.g. [17, 40]). These quality gurus and their theories were then followed by the respective Japanese experts, namely Ishikawa (e.g. [35]), Taguchi (e.g. [67]), etc., who extended the early American quality ideas and models. In the 1970–1980s, after the successes of Japanese, Americans—e.g. Crosby [14]—extended further the QM concepts.

Along with this evolution, over the years, the industrial world understood little by little the usefulness of QM standards, such as ISO 9000, ISO 22000, ISO 14000, etc., which were initiated to establish a framework on how businesses should manage their key processes. Standards can improve the organisation of enterprises, whether they manufacture products or they offer services, regardless of their size or field of activity. Moreover, they can assist businesses in clarifying their objectives and, more importantly, in avoiding expensive mistakes and nonconformities.

The positive effects of QM on companies are numerous. For example, through effective QM companies enjoy reduced costs of poor quality of products or services, increased productivity, better relationships with suppliers and customers, reduced cycle times, faster distribution of products or services to market, improved process flow, reduced waste, increased added value to customers, lower overhead costs, faster decision-making processes, operational efficiency, better working conditions, etc.

It is reasonable to conjecture that the expected benefits if QM prevails in Reverse Logistics (RL) and recovery activities, should be similar. The main objective of this book is to identify these benefits, discuss for the first time at length and in depth the role of QM in RL and examine broadly various QM issues and their interaction with the design and operation of Closed-Loop Supply Chains (CLSC).

Reverse Logistics has emerged as an important field only in the last two decades. Many years ago, Supply Chains (SC) were well-organised sequences of production processes and products, from sourcing of raw materials to disposal to the final consumers. Evidently, products are still and will always be flowing to end users, whereas at the same time a continuously increasing volume of used products has started moving backwards, namely from end users to original equipment manufacturers (OEMs) or recycling/remanufacturing companies. RL is actually the process of moving products from their typical final destination to OEMs and/or companies involved in recovery activities, in an attempt to recapture some of used products value, e.g. through resale, or for proper disposal of the item if not resalable or reusable. According to the Council of Logistics Management [65], RL is:

"The process of planning, implementing, and controlling the efficient, cost-effective flow of raw materials, in-process inventory, finished goods and related information from the point of consumption to the point of origin for the purpose of recapturing value or proper disposal."

Moreover, the following groups of activities (see also Fig. 1.1) are common in almost every product recovery system [27]:

- Collection of used products, which may include purchasing, transportation and storage activities.
- Selection—Inspection, which includes all operations determining whether returns are in fact recoverable and in which way.
- Reprocessing—Reconditioning returns, which means the actual transformation of used products into usable products again.
- Disposal is required for returns that are found to be unrecoverable due to technical or economical reasons.
- Redistribution of recovered products refers to directing them to a potential market and to physically moving them to future users. It may encompass sales (leasing, service contracts, etc.), transportation and storage activities.

The type of product recovery and the sequence of the required processing steps are often dependent on the quality condition of returns [8, 25, 29, 45], etc.. The possibilities, listed in order of the required degree of disassemble, are: repairing,

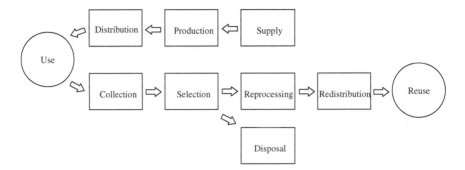

Fig. 1.1 Groups of activities in product recovery systems [27]

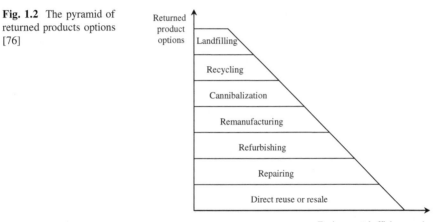

Fig. 1.2 The pyramid of returned products options [76]

refurbishing, remanufacturing, cannibalization and recycling [70]. Moreover, a returned product can be in some occasions directly resold or landfilled (Fig. 1.2). Overall, processing or handling of returned items includes:

1. Processing returned merchandise that is either damaged or seasonal, has to be restocked, is returned for salvage or for rebuild/refurbishment, has been recalled or includes items that may be an excess to the inventory.
2. Recycling of packaging, packing materials and/or containers.
3. Reconditioning for resale, recovery (refurbishing, remanufacturing etc.), disposal or donation of obsolete items.
4. Recovery of hazardous materials to prevent environmental implications.

The backward movement of used products is observed for a wide range of industrial and consumer products, including cell phones, cameras, laptops, printers, printer cartridges, tyres, power tools, pharmaceuticals, etc. For instance, in recent years the automotive industry is increasingly concerned with changing the SC to facilitate the recovery of cars that are at the end of their life [9, 10, 22]. Besides, the personal computers industry has to deal with progressively higher rates of returns.

Rather unexpectedly, the history of RL can be considered to have begun a long time ago. Its origin can be attributed to the emergence of cheap materials and advanced technologies that accompanied the industrial revolution of the 1800s. During that period, western industries followed a practice of mass production and routine throw away, with little concern for environmental matters or sustainable development [16]. Consequently, the primary motivation of RL was scarcity of resources, as the additional negative effects of such production practices did not become apparent until only in the last decades. Several years later, a similar motive, namely material shortages during the 2nd World War, boosted the remanufacturing of auto parts [60] and started a recycling trend of auto parts that continues even today.

Similarly, retail returns have their roots in the customer service policy of Montgomery Ward's, a successful mail-order business. In the late 1800s, Aaron Montgomery Ward introduced the "Satisfaction Guaranteed or Your Money Back" policy and promised to refund the purchase price to all customers dissatisfied with their purchases.

In the early 1970s, Meadows et al. [51] on behalf of the Club of Rome,[1] a non-profit global think tank, argued that there was a limit to the ongoing world growth trend. They concluded that if the prevailing trends of that period in population, pollution and resource depletion continued unchanged, the limits of population growth and worldwide economy would be reached around 2030. Then, an uncontrollable decline in both population and industrial capacity would be unavoidable. Only drastic measures for environmental protection could be suitable to change this behaviour and only under these circumstances both world population and wealth could remain at a constant level. During the following decades, this study along with several environmental disasters kept the minds of academics, politicians, industry and society in general, focused upon such environmental issues [16].

The increased social concern led to new laws, regulations and directives that changed the relationship between companies and environment. Source reduction, recycling and reuse were in fact new challenges to professionals, who had not been involved with such environmental issues in the past. Because of these new challenges, RL became a new area of concern for both industry and academics. An important milestone for RL was 1991, when Germany passed the ordinance on the "avoidance of packaging waste" that established mandatory recycling programmes for companies. This ordinance became the motivation for similar laws that were later introduced in other countries. For instance, in 1996 the respective legislation in United Kingdom required shippers and manufacturers to be responsible for the return and recycling of packaging materials. In 2001, EU took one step further by establishing a goal of 50–65 % recovery or recycling of

---

[1] http://www.clubofrome.org

packaging waste. The implication for the rest of the world was that they would have to be compliant if they wanted to do business with EU.

Another significant point in the history of RL as an academic field was the establishment of the international working group on RL, RevLog,[2] which was a cooperation of several European Universities worldwide, namely Erasmus University Rotterdam, Aristotle University of Thessaloniki, Eindhoven University of Technology, INSEAD business school, Otto-von-Guericke University Magdeburg, and University of Piraeus. The main objective of RevLog was to analyse the key issues of RL, to order them according to their impact on various industries and society and to build a framework linking these issues. An integrated approach was suggested, which considered both the traditional activities like distribution, production and inventory control, as well as the RL processes. Apart from theory building, various case studies were developed, minimising the gap between theory and practice.

In closing this introduction, it should be underlined that the fact that companies get more and more involved in reuse activities should not be attributed only to the environmental sensitivity or just to the legislation (mainly in EU, e.g. by the WEEE and RoHS directives) to produce and sell environmentally friendly products. Additionally, companies become progressively aware of the profitability of the different reuse activities. In recent years, it has been realised that the recovery systems should not be considered as a cost centre, but as a profit one [54]. RL is important for any business that is concerned not only about meeting ecological mandates and customer service, but also about the bottom line profitability.

In the remainder of this introductory chapter, we present the state-of-the-art regarding the relationship and interaction between the research areas of QM and RL. More specifically, we present a literature review[3] in order to identify the most important research articles that probe the impact of quality issues on RL. We first present the publications that do not go deeply in modelling quality issues of returned or recovered products, but simply mention some interesting and useful theories and beliefs regarding this subject (Sect. 2.1). However, in the much more extensive Sect. 2.2, we refer to publications that develop mathematical models where quality of returns or recovered products is taken into consideration either as decision variable(s) or as parameter(s) of the respective formulation.

1.2 Literature Review

In the extant literature of RL, the manner and detail in which the researchers refer to quality issues varies significantly. For example, some of them simply mention the importance of the information regarding the quality of returns and the way products were treated by users during their life, while others formulate simple or

---

[2] http://www.fbk.eur.nl/OZ/REVLOG/

[3] A much more detailed review can be found in Nikolaidis [55]

complex models trying to incorporate various quality parameters. Thereafter, we present in brief some significant articles from both these categories.

1.2.1 Publications that Simply Refer to Quality Issues of RL

Flapper [24] was among the first to consider the variable quality of returned products. In his study, he tries to give persuading answers regarding "when, where, who should collect which quantities of which returned products?". A few years later, Klausner et al. [43] considering that through repair used products are restored to "working order" and not in a "as good as new" quality condition, show that the replacement of a large part of conventional repairs with remanufacturing/reconditioning results in a higher service level for consumers and a better quality level of products.

Fleischmann et al. [26] deal with the value of information regarding the quality of returns. They admit that the knowledge of product quality as soon as possible, through modern information technology, e.g. sensor-based data recording devices, electronic data logs, etc., can result in various advantages. Simultaneously, Ferguson and Browne [22] refer to the utilisation-related information—such as service life, amount of usage, etc.—which can reveal the use of a product over its entire life. This type of information is necessary in order to assess the quality of any returned product and determine its potential remaining value.

A few years later, Blackburn et al. [7] examine a number of companies that extract value from product returns and mention that the variety and quality of returns are already determined upon their receipt. However, the latter may not be observable without examination and testing of returned products. Either way, it is a key decision for a company to evaluate its returns immediately in order to determine their quality level. At the same time, Toffel [71] deals with the strategic management of product recovery and focuses on factors that motivate OEMs to engage in EOL product recovery activities. He also refers to various quality issues of product recovery.

Finally, very recently, Wong et al. [78] provide empirical evidence on the performance value of asset recovery. More specifically, they highlight the potential value of the product recovery process in a CLSC and shed light on its performance implications in terms of financial performance and product quality.

1.2.2 Publications with Mathematical Models Regarding Quality Issues of RL

By focusing on one product type and taking into account technical, commercial and ecological criteria, Krikke et al. [45] develop a Dynamic Programming model which determines the optimal product recovery and disposal strategy (see also

Fig. 1.2). Their model tries to maximise the expected net profit, considering also the uncertainty on the aforementioned criteria due to lack of information about the quality condition of a product and its components. Almost a decade later, Jun et al. [39] develop a multi-objective algorithm that enables the selection of the best option of parts or components of EOL products in order to maximise the recovery value of an EOL product. The most representative criteria to evaluate this type of value are recovery cost and recovery quality. Regarding the latter, they consider two types of EOL product quality: quality before and after recovery. The problem of optimising product recovery options is also investigated by Lamsali and Liu [48]. They develop a linear programming model that maximises the profit in order to find the optimal classification of returns in different quality classes, according to certain recovery options. Wadhwa et al. [77] have almost the same ambition: to determine the best alternative for reprocessing EOL products bringing a fuzzy decision-making method and RL together. More specifically, the decision analysis they use allows decision makers to rank the order of the alternatives based on the results of this analysis. They take into consideration that the nature of RL decisions is usually multidimensional, interdisciplinary and complex due to uncertainties in quality, quantity and timing of returns.[4]

In the mean time, Guide et al. [30] focus on the CLSC of Hewlett-Packard for returned notebooks and PCs in Europe. They develop a number of decision models in order to help the decision-making process of the company. Specifically, their multiperiod linear programming models explore alternative scenarios with the objective of profit maximisation and investigate a range of values for the returned notebooks quality. In the next year, Pochampally and Gupta [58], alternatively to the traditional approach for six-sigma investigation, develop a new and less confusing formula, which they apply to select potential recovery facilities in a region where a CLSC is to be designed. Recently, Van Wassenhove and Zikopoulos [76] present the most important decisions that should be made regarding the establishment of the assessment and classification of used products according to their quality condition. In particular, they pay special attention to cases of firms which either maintain multiple production stations, each one dedicated to a specific product quality level, or perform different recovery operations according to the condition of each returned unit.

The rest of publications regarding QM issues in RL that are also presented in this literature review are grouped according to significant features of RL, such as: (used) product acquisition, collection of returns, inventory management, disassembly, production planning and control, information technology for RL, pricing (e.g. of remanufactured products) and environmentally friendly product design. Moreover, several of the following subsections give emphasis to remanufacturing and the majority of the aforementioned features, as remanufacturing is in practice one of the most popular forms of reprocessing of returned products.

---

[4] Several authors mention these types of uncertainties; for instance Krikke et al. [45, 46], Zhou et al. [81], Ketzenberg et al. [41] etc.

1.2.2.1 Product Acquisition/Collection

Some commonly studied RL issues concerning product acquisition include the implementation of different forms of financial incentives, their impact on the performance of RL activities and the selection of optimal return policies. Usually, these issues are examined in relation to the quality of returned units, among other things. For instance, Guide and Van Wassenhove [32] present the real case of a cell phone remanufacturing company (ReCellular) to illustrate the implementation of a quality-dependent incentives policy. Despite the fact that there is always quality uncertainty in return flows, it is possible for a firm to manage the quality of product returns by offering financial incentives. Therefore, predetermined acquisition prices may be offered for products in different quality levels. Two years later, Guide et al. [31] refer again to ReCellular, which acquires used phones in different quality conditions, however it remanufactures them to a single quality level and sells them at a specific price. They consider that quality, quantity and timing of return flows can be controlled by the offered acquisition prices.

Along a similar research line, Aras and Aksen [1] and Aras et al. [2] determine optimal incentive values under a quality-dependent incentive policy. More specifically, Aras and Aksen [1] propose a new facility location-allocation model which helps a company to determine the optimal incentives that should be offered to product holders, as well as the optimal number and locations of collection centres. In order to incorporate the willingness of product holders in their return process model, Aras and Aksen [1] consider that an important factor that affects this willingness is the amount of financial incentives offered to them. Aras et al. [2] propose a similar facility location-allocation model which might help a company to face the problem of deciding the best incentive values offered to product holders depending on the quality condition of the items they possess, as well as the best locations of collection centres.

In contrast with all the above researchers, Mitra [53] assumes that a remanufacturer has no control on quality, quantity and timing of returns.

Finally, Dobos and Richter [19] (see additional information in Sect. 2.2.8) consider two strategies to manage the collection of used items: either to repurchase only the reusable products or to buy back all items and then investigate their serviceability. The question they try to answer in their study is which of the two strategies minimises the relevant costs.

1.2.2.2 Disassembly

As far as disassembly is concerned, Fleischmann [25] mentions that the recovery of valuable parts by dismantling used products is perceived as a cheap yet uncertain supply source. Quality is a major issue in this context; despite the fact that a used machine that is returned by a customer is not always defective, its quality certification may require expensive inspection and testing. He also presents the case of IBM which is careful not to corrupt its quality standards by embedding used equipment into the spare parts circuit. A few years later, Johar and Gupta [38] analyse the inventory control problem of returned EOL products in the disassembly context. To solve this problem, they introduce a multiperiod dynamic programming model, the goal of which is to optimise the total expected profit. A probabilistic approach is used where the EOL products have different quality levels depending on the environments they have been operating in. Moreover, different quality levels for every disassembled component are considered.

Tang et al. [69] explain that disassembly is manual, semiautomatic and labour intensive. In their article, as well as the one of Tang and Zhou [68], several factors of the human intervention in the disassembly process—such as the quality of disassembled subassemblies and components—are studied, using a fuzzy attributed Petri net model. Employing again the technique of fuzzy logic, Tripathi et al. [72] attempt to maximise the net profits associated with the product/part recovery, through the minimisation of the disassembly cost in a real world environment. Unfortunately, in this environment the knowledge related to the quality of returned products and their constituent parts is quite vague.

1.2.2.3 Production Planning/Control

Kleber et al. [44] determine the optimal control policy between production, remanufacturing and disposal, assuming a linear cost model and different costs for the variety of remanufacturing and production processes. They consider a single deterministic return stream where all returns have the same quality level (!) and can be split and remanufactured to serve multiple demand streams, e.g. for recovered products of different qualities. Similarly, Aras et al. [4] develop a simulation model of a hybrid system where the quality of returns is uncertain, in order to compare two alternative strategies that use either manufacturing or remanufacturing as the primary means of satisfying customer demand. Another simulation model is constructed by Behret and Korugan [6] who analyse the effect of uncertainties in quality of returns, considering the classification of returned products in three quality levels. They find an effective policy that balances both manufacturing and remanufacturing throughputs with demand.

Very recently Das and Chowdhury [15] have introduced a modular product design for RL planning that minimises recovery costs and improves the overall production performance of a CLSC that involves new and recovered products. More specifically, they develop a mixed integer programming model in order to formulate the overall planning process required to maximise profit by considering the collection of returned products, the recovery of modules and the proportion of the acquired product mix at different quality levels.

1.2.2.4 Information Technology for RL

Regarding information technology in the area of RL, Klausner et al. [42] first clarify that to make product recovery economically attractive, certain information on the history of products use is needed. To this end, they propose the

Fig. 1.3 RFID tag on an
automotive component
(reproduced from [57])

"Information System for Product Recovery", whose essential component is an electronic device, the so-called "Electronic Data Log". This device, which is integrated into products, gathers information on their actual usage. It works by measuring data closely connected with the quality degradation of both products and their components, over the products' entire lifetime. Almost a decade later, Ustundag et al. [73] develop a simulation model for RFID tags. Using this model, they evaluate the influence of various factors, such as the quality of tags, on the total system cost which is determined by purchasing and RL costs of tags. They show that the higher quality tags lead to lower system costs. Along the same line of research, Parlikad and McFarlane [57] focus on automotive parts recovery and illustrate how decision-making during product recovery, can be improved by collecting critical usage data during a vehicle's life cycle. They explain that the quality of an automotive component depends on a number of factors such as its age, reliability, usage, maintenance, severity of environment, etc. In order to assess this quality, updated information arising from all phases of the component's life cycle is required. This information can be collected through RFID tags (Fig. 1.3).

1.2.2.5 Pricing

Vadde et al. [74] present four pricing models for various grades of reusable and recyclable components when product recovery facilities either passively accept or proactively acquire single and multitype product returns. More specifically, they consider the demand for high quality recyclable, scrap grade reusable and scrap quality recyclable components, as well as remanufactured components as factors that decide the acquisition quantity. The decision variables in their optimization problem are the sale prices of the aforementioned components, as well as the price of damaged discarded products. Similarly, Mitra [53], presented initially in Sect. 2.2.1) extends the model of Guide et al. [31] by considering that there might be more than one quality level of recovered products, the prices of which are decision variables.

1.2.2.6 Remanufacturing—Product Acquisition/Collection

Robotis et al. [59] study the use of remanufacturing as a tool to serve secondary markets and contribute to the understanding of interactions between procurement and remanufacturing decisions. They model the case where a reseller acquires used products of an old technology from an advanced market; their quality is not fixed and it is considered to be a decision variable and can be changed through remanufacturing. Bakal and Akcali [5] motivated by an application in automotive parts remanufacturing industry, develop a model that incorporates randomness in the supply process and lets the expected quality of returned products be dependent on the price offered by the remanufacturer. Galbreth and Blackburn [28] derive optimal sorting and acquisition policies in the presence of used product condition variability: from slightly used products with only minor cosmetic faults to significantly damaged ones, requiring extensive rework. They use a total cost model that incorporates the variable condition of used products that remanufacturers usually face.

1.2.2.7 Remanufacturing—Used Product Disposition
(Sorting, Testing, Grading)

Souza et al. [64] investigate the impact of inaccuracies in grading the quality of product returns (see additional information in Sect. 2.2.9), while Aras et al. [3] formulate a continuous time Markov chain model to analyse the conditions under which quality-based categorization of returned products leads to cost savings in hybrid manufacturing/remanufacturing systems. They identify four cases where the quality-based categorization is most cost-effective: when quality difference between returned products is high, the quality of both high- and low-quality returns is low, return rate is high and demand rate is low.

Zikopoulos and Tagaras [83] examine the usefulness of simple sorting procedures for a single-period setting, which are used in order to test the quality of returns prior to disassembly and remanufacturing processes. In particular, they study the impact of using such quick but inaccurate sorting techniques on the profitability of remanufacturing operations. The part of the work in Zikopoulos and Tagaras [83] that examines briefly the infinite-horizon problem for a single collection site is extended by Tagaras and Zikopoulos [66]. Regarding the uncertainty of returned units quality, they mention that there are two significant issues: (i) the unknown number of remanufacturable units in a lot of returned items, while (ii) even if that number can be estimated accurately, there is still the need to identify the specific remanufacturable units. To this end, Tagaras and Zikopoulos [66] study the feasibility of establishing an early sorting operation in a CLSC with multiple collection centres and a central remanufacturing facility. They formulate the expected profit functions and specify the optimal procurement decisions for three system settings: (i) without sorting, (ii) with central sorting and (iii) with local sorting.

Finally, Van Wassenhove and Zikopoulos [75] examine thoroughly the issue of overestimation of returned products quality and its effect on the profitability of a remanufacturer. More specifically, they examine how misclassification of used products affects the loss of a remanufacturer and, consequently, to what extent a firm should undertake some actions in order to improve the classification accuracy. An interesting conclusion they raise is that the economical impact of poor classification is primarily affected by the extent of quality overestimation.

1.2.2.8 Remanufacturing—Inventory Management

The work of Dobos and Richter [19] focuses on remanufacturing and refers to inventory management. They extend a previous model of their own [18] by relaxing the assumption of perfect quality of returned units. They investigate a production-recycling model with quality consideration, assuming that the quality of returned and collected products is not always suitable for further exploitation, i.e. not all used items can be reused. However, the proportion of serviceable items is known with certainty. Jaber and El Saadany [36], and El Saadany and Jaber [21] extend the same model of Dobos and Richter [18], by assuming that the collection rate of used items is dependent on the purchasing price and the quality level of returns. Especially in the latter, this is done by incorporating a price-quality demand function which presents demand as a decreasing and increasing exponential function of price and quality, respectively. Moreover, El Saadany and Jaber [21] mention that although several researchers have pointed out the need to differentiate returned units according to their quality [6, 7, 26], etc., there has not been any work that models the collection rate of used items as price and quality dependent.

1.2.2.9 Remanufacturing—Production Planning/Control

In the area of production planning and control, Souza et al. [64] model a remanufacturing facility as a queuing network where used products are sorted and graded at different remanufacturing stations into three quality classes. The remanufacturer can either sell returned items of different quality grades on a graded as-is basis or remanufacture them and improve their quality. Their model determines the firm's decision to remanufacture an optimal product mix over the long run by maximising profits, while maintaining desired flow times. The optimal remanufacturing policy provides the percentage of each quality grade to be remanufactured at each station. Inderfurth [34] uses numerical analysis to investigate the effect of uncertainties related with quantity of returns, quality and demand, on product recovery activities. For this purpose, he introduces a model of a hybrid production and remanufacturing system which includes uncertainty, by describing demands for new products and returns of used products with different quality levels, as stochastic processes.

Jayaraman [37] presents an analytical approach for CLSC with product recovery and reuse, which consists of a mathematical programming model for aggregate production planning and control. Jayaraman's model is designed to aid operational decision makers in an intermediate to long-range planning environment. Key decisions include the number of units with a nominal quality level that should be disassembled, disposed, remanufactured and acquired in a given time period, as well as the inventory of modules and cores that remain at the end of a given time period. Zikopoulos and Tagaras [82] refine the research of Inderfurth [34] by investigating in a single-period setting the impact of uncertainty in the quality of returned products on the profitability of specific reuse activities. They consider a CLSC with a single refurbishing location supplied by two collection sites. Sorting and refurbishing take place at the aforementioned location. Additionally, the quality of returned products which are available at the collection sites is uncertain and may only be revealed when these products are transported to the refurbishing location.

In a research article similar to the one of Aras et al. [3], Ferguson et al. [23] study a production-planning problem for remanufacturing where the recoverable products have heterogeneous quality levels. They consider a fuzzy setting for the quality of returns, while the cost of remanufacturing depends on this quality in the following way: the worse the quality of returns, the higher the cost to remanufacture them (considering the higher amount of time and materials for this remanufacturing). They formulate a stochastic dynamic programming model which must be implemented in a rolling horizon, as stochastic demand and returns are realised periodically.

1.2.2.10 Environmentally Friendly Product Design

Ilgin and Gupta [33] review the literature on environmentally conscious manufacturing as well as product recovery, by systematically investigating and classifying lots of published articles. An interesting part of their review refers to Quality Function Deployment (QFD)-based methodologies that have been developed in order to achieve the so-called "design for environment" (DFE), i.e. the design of products so that their potential environmental impact throughout their life cycle is minimised. Moreover, this type of product design is also quality related and recovery conscious, and thus it is interesting as far as our review is concerned

Cristofari et al. [13] introduce an innovative design methodology called Green QFD which considers quality requirements, environmental impact, and production costs at the design phase. More specifically, it integrates QFD and Life Cycle Assessment (LCA) into a powerful tool that can help design teams document the technical requirements for a product concept, while assessing the environmental impacts associated with that concept. This tool has been further improved in the Green QFD-II methodology developed by Zhang et al. [80]. Their methodology combines LCA and Life Cycle Costing (LCC) into QFD matrices and provides a mechanism that deploys all requirements, namely quality, environmental and costing requirements throughout the entire product development process. Green QFD-III proposed by Mehta and Wang [52] utilise a method (Eco-Indicator '99) for quantifying the environmental impact of the product and simplifies the detailed LCA and complex product comparison algorithm of Green QFD-II. Finally, Dong et al. [20] develop Green QFD-IV and improve its predecessor by using a fuzzy multiattribute utility method to estimate the life cycle cost.

Santos-Reyes and Lawlor-Wright [62] develop a four-phase product design methodology that includes first the understanding of environmental problems, and then its transformation (using the Analytic Hierarchy Process—AHP), manipulation (using QFD) and evaluation. Masui et al. [50] suggest another four-phase methodology for DFE, which incorporates environmental aspects into QFD to handle both the environmental and traditional product quality requirements. This methodology is intended to be used in the early stages of new product design. A few years later, Sakao [61] extends the method of Masui et al. [50] by employing LCA and theory of inventive problem solving.

Madu et al. [49] propose a method that integrates stakeholders into an environmentally conscious design for products and/or processes. First, AHP is used to prioritise customer requirements. Then, QFD is used to match design requirements to customer ones. Finally, a cost-effective design strategy is developed using Taguchi experimental design and loss function. Madu et al. [49] illustrate the proposed framework in a paper recycling problem. Kuo and Wu [47] apply QFD to translate customer needs into six categories of environmentally technical measures. Then, they determine the best design alternative based on the product's life cycle, i.e. raw material, manufacturing, assembly, disassembly, transportation, customer usage and disposal. Bovea and Wang [11] introduce an approach for identifying environmental improvement options by taking into account customer preferences. They apply LCA to evaluate the environmental profile of a product, while a fuzzy approach based on the House of Quality in the QFD methodology provides a more quantitative method for evaluating the imprecision of human qualitative assessments. A few years later, the same researchers [12] propose a redesign approach that allows integrating environmental requirements into product development. Their methodology establishes a relationship between QFD, LCA, LCC and contingent valuation techniques for evaluating the customer, environmental, cost requirements and customer willingness-to-pay, respectively. Recently, Yüksel [79] evaluates the application of QFD methodology and House of Quality in order to design products that are suitable for remanufacture.

In their slightly different paper Park and Tahara [56] mention that it is possible to design a product that is environmentally friendly, while still maintain a high level of quality and consumer satisfaction. They simultaneously consider the quality, environmental and customer satisfaction related aspects of products by using Producer-Based and Consumer-Based eco-efficiency. The first is used to identify the key issues of a product as related to product quality and environmental impact whereas the second methodology is used to identify consumer satisfaction and environmental impact related product characteristics.

1.3 Conclusions

The main conclusion that can be drawn from the previously presented review is that the majority of researchers simply consider the uncertainty and the variability of the returned units' quality in a superficial level, i.e. they do not actually try to find ways to either influence or improve it. That improvement could result either through providing incentives to customers to return the products they use sooner and/or in a better quality level or through the implementation of modern selling mechanisms such as leasing. The improvement of the quality of returns could also be done through the application of Acceptance Sampling which seems to be the most suitable Statistical Quality Control technique for RL and CLSC. The implementation of Acceptance Sampling would permit any company involved in reuse activities to have better knowledge of the returned products quality. Thus, decisions about the acquisition and recovery of batches of returns would become much more accurate; they would be based on recent information and not on old and probably unreliable data.

Moreover, RFID and more generally the information technology could provide the means to capture useful and accurate information regarding the way a product was used during its life cycle. Although this specific research area is very recent, it is expected to bring some impressive results in the near future.

References

1. Aras N, Aksen D (2008) Locating collection centers for distance- and incentive-dependent returns. Int J Prod Econ 111(2):316–333
2. Aras N, Aksen D, Tanugur AG (2008) Locating collection centers for incentive dependent returns under a pick-up policy with capacitated vehicles. Eur J Oper Res 191(3):1223–1240
3. Aras N, Boyaci T, Verter V (2004) The effect of categorizing returned products in remanufacturing. IIE Trans 36(4):319–331
4. Aras N, Verter V, Boyaci T (2006) Coordination and priority decisions in hybrid manufacturing/remanufacturing systems. Prod Oper Manag 15(4):528–543
4. Aras N, Verter V, Boyaci T (2006) Coordination and priority decisions in hybrid manufacturing/remanufacturing systems. Prod Oper Manag 15(4):528–543
5. Bakal IS, Akcali E (2006) Effects of random yield in remanufacturing with price sensitive supply and demand. Prod Oper Manag 15(3):407–420
6. Behret H, Korugan A (2009) Performance analysis of a hybrid system under quality impact of returns. Comput Ind Eng 56(2):507–520
7. Blackburn JD, Guide VD, Souza GC, Van Wassenhove LV (2004) Reverse supply chains for commercial returns. Calif Manag Rev 46(2):6–22
8. Bloemhof-Ruwaard JM, Fleischmann M, van Nunen JAEE (1999) Reviewing distribution issues in reverse logistics. New Trends Distrib Logist LNEMS 480:23–44
9. Boon JE, Isaacs JA, Gupta SM (2000) Economic impact of aluminum-intensive vehicles on the u.s. automotive recycling infrastructure. J Ind Ecol 4(2):117–134
10. Boon JE, Isaacs JA, Gupta SM (2003) End-of-life infrastructure economics for "clean vehicles" in the United States. J Ind Ecol 7(1):25–45
11. Bovea MD, Wang B (2003) Identifying environmental improvement options by combining life cycle assessment and fuzzy set theory. Int J Prod Res 41(3):593–609

12. Bovea MD, Wang B (2007) Redesign methodology for developing environmentally conscious products. Int J Prod Res 45(18):4057–4072
13. Cristofari M, Deshmukh A,Wang B (1996) Green quality function deployment. Proc of the 4th international conference on environmentally conscious design and manufacturing, Cleveland, 23–25 July 1996, pp. 297–304
14. Crosby P (1979) *Quality is free*. McGraw-Hill, New York
15. Das K, Chowdhury AH (2012) Designing a reverse logistics network for optimal collection, recovery and quality-based product-mix planning. Int J Prod Econ 135(1):203–221
16. De Brito MP, Dekker R (2002) Reverse logistics-a framework econometric. Institute report EI 2002–38. Erasmus University, Rotterdam
17. Deming WE (1944) A view of the statistical method. Acc Rev 19(3):254–260
18. Dobos I, Richter K (2004) An extended production/recycling model with stationary demand and return rates. Int J Prod Econ 90(3):311–323
19. Dobos I, Richter K (2006) A production/recycling model with quality consideration. Int J Prod Econ 104(2):571–579
20. Dong C, Zhang C, Wang B (2003) Integration of green quality function deployment and fuzzy multi-attribute utility theory-based cost estimation for environmentally conscious product development. Int J Environ Conscious Des Manuf 11(1):12–28
21. El Saadany AMA, Jaber MY (2010) A production/remanufacturing inventory model with price and quality dependant return rate. Comput Ind Eng 58(3):352–362
22. Ferguson N, Browne J (2001) Issues in end-of-life product recovery and reverse logistics. Prod Plan Control 12(5):534–547
23. Ferguson M, Guide VDR Jr, Koca E, Souza GC (2009) The value of quality grading in remanufacturing. Prod Oper Manag 18(3):300–314
24. Flapper SDP (1993) On the logistics of recycling. An introduction, Eindhoven University of Technology, Technical report TUE/BDK/LBS/93-16
25. Fleischmann M (2000) Quantitative models for reverse logistics, PhD thesis, Rotterdam, Erasmus University
26. Fleischmann M, Beullens P, Bloemhof-Ruwaard JM, Wassenhove LNV (2001) The impact of product recovery on logistics network design. Prod Oper Manag 10(2):156–173
27. Fleischmann M, Krikke HR, Dekker R, Flapper SD (2000) A characterization of logistics networks for product recovery. Omega 28(6):653–666
28. Galbreth MR, Blackburn JD (2006) Optimal acquisition and sorting policies for remanufacturing. Prod Oper Manag 15(3):384–392
29. Guide VDR Jr, Jayaraman V, Srivastava R, Benton WC (2000) Supply chain management for recoverable manufacturing systems. Interfaces 30(3):125–142
30. Guide VDR Jr, Muyldermans L, Van Wassenhove LN (2005) Hewlett-Packard company unlocks the value potential from time sensitive returns. Interfaces 35(4):281–293
31. Guide VDR Jr, Teunter RH, Van Wassenhove LN (2003) Matching demand and supply to maximize profits from remanufacturing. Manuf Serv Oper Manag 5(4):303–316
32. Guide VDR, Van Wassenhove LN (2001) Managing product returns for remanufacturing. Prod Oper Manag 10(2):142–155
33. Ilgin MA, Gupta SM (2010) Environmentally conscious manufacturing and product recovery (ECMPRO): a review of the state of the art. J Environ Manag 91(3):563–591
34. Inderfurth K (2005) Impact of uncertainties on recovery behavior in a remanufacturing environment. Int J Phys Distrib Logist Manag 35(5):318–336
35. Ishikawa K (1985) *What is total quality control? The Japanese way*, D. J. Lu (translation). Prentice Hall, New Jersey
36. Jaber MY, El Saadany AMA (2009) The production, remanufacture and waste disposal model with lost sales. Int J Prod Econ 120(1):115–124
37. Jayaraman V (2006) Production planning for closed-loop supply chains with product recovery and reuse: An analytical approach. Int J Prod Res 44(5):981–998
38. Johar BO, Gupta SM (2008) Analysis of inventory management in reverse supply chain using stochastic dynamic programming model. Proc ASME Int Mech Eng Cong Expo 8:945–953

39. Jun H-B, Cusin M, Kiritsis D, Xirouchakis P (2007) A multi-objective evolutionary algorithm for EOL product recovery optimization: turbocharger case study. Int J Prod Res 45(19): 4573–4594
40. Juran JM, Gryna FM (1951) *Juran's quality control handbook*. Mcgraw-Hill, New York
41. Ketzenberg ME, Van der Laan E, Teunter RH (2006) The value of information in closed loop supply chains. Prod Oper Manag 15(3):393–406
42. Klausner M, Grimm W, Hendrickson C, Horvath A (1998) Sensor-based data recording of use conditions for product takeback. Proc IEEE Int Sympos Electron Environ, Chicago, pp 138–143
43. Klausner M, Grimm WM, Horvath A (1999) Integrating product take-back and technical service. Proc IEEE Int Sympos Electron Environ, Danvers, pp 48–53
44. Kleber R, Minner S, Kiesmuller G (2002) A continuous time inventory model for a product recovery system with multiple options. Int J Prod Econ 79(2):121–141
45. Krikke HR, Van Harten A, Schuur PC (1998) On a medium term product recovery and disposal strategy for durable assembly products. Int J Prod Res 36:111–139
46. Krikke HR, Van Harten A, Schuur PC (1999) Business case Oce: reverse logistic network re-design for copiers. OR Spektrum 23(3):381–409
47. Kuo TC, Wu HH (2003) Green products development by applying grey relational analysis and green quality function deployment. Int J Fuzzy Syst 5(4):229–238
48. Lamsali H, Liu J (2008) Optimizing the selection of product recovery options, industrial engineering and engineering management, IEEM 2008. IEEE international conference issue, pp. 1704–1708
49. Madu CN, Kuei C, Madu IE (2002) A hierarchic metric approach for integration of green issues in manufacturing: a paper recycling application. J Environ Manag 64(3):261–272
50. Masui K, Sakao T, Kobayashi M, Inaba A (2003) Applying quality function deployment to environmentally conscious design. Int J Qual Reliab Manag 20(1):90–106
51. Meadows D, Meadows D, Randers J, Behrens WW III (1972) *The limits to growth*. Universe Books, New York
52. Mehta C, Wang B (2001) Green quality function deployment III: a methodology for developing environmentally conscious products. J Des Manuf Autom 4(1):1–16
53. Mitra S (2007) Revenue management for remanufactured products. Omega 35(5):553–562
54. Nikolaidis Y (2009) A modelling framework for the acquisition and remanufacturing of used products. Int J Sustain Eng 2(3):154–170
55. Nikolaidis Y (2012) Quality management issues in reverse logistics: reviewing the state of the art, working paper. University of Macedonia, Naoussa, Greece
56. Park PJ, Tahara K (2008) Quantifying producer and consumer-based eco-efficiencies for the identification of key ecodesign issues. J Clean Prod 16(1):95–104
57. Parlikad AK, McFarlane D (2010) Quantifying the impact of AIDC technologies for vehicle component recovery. Comput Ind Eng 59(2):296–307
58. Pochampally KK, Gupta SM (2006) Total quality management (TQM) in a reverse supply Chain. Proc of the SPIE International Conference on Environmentally Conscious Manufacturing VI, Boston, Massachusetts, pp 139–148
59. Robotis A, Bhattacharya S, Van Wassenhove LN (2005) The effect of remanufacturing on procurement decisions for resellers in secondary markets. Eur J Oper Res 163(3):688–705
60. Rogers D, Tibben-Lembke RS (1999) *Going backwards: reverse logistics trends and practices*. RLEC Press, Pittsburgh
61. Sakao T (2007) A QFD-centred design methodology for environmentally conscious product design. Int J Prod Res 45(18):4143–4162
62. Santos-Reyes DE, Lawlor-Wright T (2001) A design for the environment methodology to support an environmental management system. Integr Manuf Syst 12(5):323–332
63. Shewhart WA (1924) Some applications of statistical methods to the analysis of physical and engineering Data. Bell Syst Tech J 3(1):43–87
64. Souza G, Ketzenberg M, Guide VDR (2002) Capacitated remanufacturing with service level constraints. Prod Oper Manag 11(2):231–248

65. Stock JR (1992) Reverse logistics. Council of Logistics Management, Oak Brook
66. Tagaras G, Zikopoulos C (2008) Optimal location and value of timely sorting of used items in a remanufacturing supply chain with multiple collection sites. Int J Prod Econ 115(2): 424–432
67. Taguchi G, Wu Y (1980) *Introduction to off-line quality control*. Central Japan Quality Association, Nagoya, Japan
68. Tang Y, Zhou M (2008) Human-in-the-loop disassembly modeling and planning. In: Gupta S, Lambert A (eds) *Environment conscious manufacturing*. CRC Press, Boca Raton
69. Tang Y, Zhou M, Gao M (2006) Fuzzy-petri-net based disassembly planning considering human factors. IEEE Trans Syst Man Cybern 36(4):718–726
70. Thierry M, Salomon M, van Nunen J, van Wassenhove LN (1995) Strategic issues in product recovery management. Calif Manag Rev 37(2):114–135
71. Toffel MW (2004) Strategic management of product recovery. Calif Manag Rev 46(2):120
72. Tripathi M, Agrawal S, Pandey MK, Shankar R, Tiwari MK (2009) Real world disassembly modeling and sequencing problem: optimization by algorithm of self-guided ants (ASGA). Robot Computer-Integrated Manuf 25(3):483–496
73. Ustundag A, Baysan S, Çevikcan E (2007) A conceptual framework for economic analysis of REID reverse logistics via simulation, 2007 1st Annual RFID Eurasia, art. no. 4368144
74. Vadde S, Kamarthi SV, Gupta SM (2007) Optimal pricing of reusable and recyclable components under alternative product acquisition mechanisms. Int J Prod Res 45(18–19): 4621–4652
75. Van Wassenhove LN, Zikopoulos C (2010) On the effect of quality overestimation in remanufacturing. Int J Prod Res 48(18):5263–5280
76. Van Wassenhove LN, Zikopoulos C (2011) Quality in reverse. Ind Eng 43(3):41–45
77. Wadhwa S, Madaan J, Chan FTS (2009) Flexible decision modelling of reverse logistics system: a value adding MCDM approach for alternative selection. Robot Computer-Integrated Manuf 25(2):460–469
78. Wong CWY, Lai K, Cheng TCE, Venus Lun YH (2012) The roles of stakeholder support and procedure-oriented management on asset recovery. Int J Prod Econ 135(2):584–594
79. Yüksel H (2010) Design of automobile engines for remanufacture with quality function deployment. Int J Sustain Eng 3(3):170–180
80. Zhang Y, Wang HP, Zhang C (1999) Green QFD-II: a life cycle approach for environmentally conscious manufacturing by integrating LCA and LCC into QFD matrices. Int J Prod Res 37(5):1075–1091
81. Zhou L, Disney SM, Lalwani CS, Wu H (2004) Reverse logistics: a study of bullwhip in continuous time. Proc World Cong Intell Control Autom (WCICA) 4:3539–3542
82. Zikopoulos C, Tagaras G (2007) Impact of uncertainty in the quality of returns on the profitability of a single-period refurbishing operation. Eur J Oper Res 182(1):205–225
83. Zikopoulos C, Tagaras G (2008) On the attractiveness of sorting before disassembly in remanufacturing. IIE Trans 40(3):313–323

Chapter 2
A Quality Framework in Closed Loop Supply Chains: Opportunities for Value Creation

Umut Çorbacıoğlu and Erwin A. van der Laan

Abstract Quality issues and "uncertainties" are encountered in almost every aspect of closed loop supply chains (CLSCs). In this chapter, we analyze the CLSC processes with a focus on quality. We find that quality of returned products is the major source of uncertainty and thus a major determinant of value in CLSCs. However, we observe that there is a gap in the literature when it comes to identifying and properly defining all relevant quality dimensions and the ways in which they affect CLSC processes. In our chapter, we start with an investigation of existing definitions of quality, link them to the different stages of CLSCs, and propose a new framework summarizing the integration of quality within CLSC processes. We also relate our framework to other frameworks that can be found in the literature and show how the former may help to improve value creation in CLSCs.

2.1 Introduction

A closed loop supply chain (CLSC) is defined in [1] as "...a system to maximize value creation over the entire life cycle of a product with dynamic recovery of value from different types and volumes of returns over time". Quality issues are encountered in almost every aspect of CLSCs: as a driver of return flows, as a marketing challenge for recovered products, and, more importantly, as an

U. Çorbacıoğlu
Quintiq Applications B.V., Bruistensingel 500, Den Bosch, The Netherlands
e-mail: Umut.Corbacioglu@quintiq.com

E. A. van der Laan (✉)
Rotterdam School of Management, Erasmus University Rotterdam, 1738Rotterdam, The Netherlands
e-mail: ELaan@rsm.nl

Y. Nikolaidis (ed.), *Quality Management in Reverse Logistics*,
DOI: 10.1007/978-1-4471-4537-0_2, © Springer-Verlag London 2013

uncertainty of input to the recovery process. When referring to quality in the context of product recovery several terms such as "reusability", "condition of items", "remaining life", "residual functionality", and "remaining value" are encountered. For instance, Kumer et al. [2] claim that quality is one of the determinants of value creation, but they do not define quality and do not explain how quality affects value creation; Visich et al. [3] claim that the use of RFID may enhance value creation, in particular through identifying the quality levels of a product return, but quality is not defined. Zuidwijk and Krikke [4] investigate possible strategic responses to electric and electronic equipment (EEE) returns and define several scenarios with respect to product returns' quality ("low", "high" etc.), but do not define quality. Moreover, they mention that product returns with high quality may be brought to "as good as new" quality.

These numerous, yet often implicit, references to quality point to the necessity to properly define and structure all relevant quality dimensions of CLSCs. Only when these dimensions and their consequences are made explicit, specific targets and actions can be defined to improve supply chain performance. Hence, the main contribution of this chapter is the construction of a quality framework for CLSCs that identifies all relevant quality dimensions in the CLSC, defines them explicitly, and links these dimensions to various processes in the CLSC. Hence, the framework enables the analysis of CLSCs through a quality lens and in a structured way, as well as the discovery of new ways of value creation. Also, we hope to lay a foundation for future research, especially for the Operations Management discipline.

In the literature, several frameworks have been presented to analyze product recovery systems (e.g., [5–8]). However, to the best of our knowledge there has not been any study that developed a framework specifically from a quality point of view. To construct our framework, we start (Sect. 2.2) with discussing (returns') uncertainty and its effects on CLSC-related processes, to establish quality as the key source of uncertainty. After that, in Sect. 2.3 we discuss how quality is traditionally defined, how the concept of "customer value" (mentioned e.g., in [9]) is built on these definitions, and how quality management principles follow from the concept of product value. Further, we employ the customer value definition to define the "return value". In Sect. 2.4, we focus on how the traditional quality definitions apply in product recovery settings and identify which of those definitions are applicable at what stages of the CLSC. Based on this mapping we construct our framework. Then, having identified the traditional definitions of quality that apply to product recovery, in Sect. 2.5 we focus on the quality management approaches and their applicability in product recovery setting. In the last but one section, we relate our quality-based framework to earlier ones that can be found in the literature to establish a link between planning and management of product recovery and our quality framework. Finally, in Sect. 2.6 we present potential issues of future research.

2.2 Uncertainty in Product Recovery Environments

Return flows are characterized by high levels of uncertainty. When describing them, several researchers (e.g., [10, 11]) have identified uncertainties with respect to quantity, timing, and quality of returns. In this section we argue that quality is the major source of uncertainty in CLSCs. We further argue that it encompasses timing and quantity uncertainties after the receipt of the recoverables due to potential yield loss.

Categories of recovered products may differ with respect to their coordinates on the dimensions of quantity, timing, and quality. In this regard, Guide [8] argues that in recovery environments uncertainties in timing is positively correlated with uncertainties in quantity, but negatively correlated with quality uncertainty. Using Guide's examples, for remanufacturing of jet engines timing is highly predictable, whereas the condition of components is variable. On the other hand, for remanufacturing of single use cameras quantity and timing is highly uncertain, whereas quality variance is limited. Nevertheless, virtually in all applications all three sources of uncertainty are present to some degree and complicate the decision-making processes.

Timing and quantity uncertainty are linked to the product's life cycle, rate of technological change, and the willingness of end-users to return the product [12]. The condition of recoverables, on the other hand, depends on factors such as the age of product, the customer use pattern, and the nature of product (e.g., mechanical or electronic).

Along the CLSC, quantity and timing uncertainties are partially resolved at the same stage, that is, upon receipt of the recoverable items. Le Blanc [13] introduced the concept of disposer decoupling point (DDP) for CLSCs based on the customer decoupling point (CDP) concept of traditional forward supply chains. The DDP corresponds to the point in the CLSC where the "disposition route" can be actively managed, i.e., where the decisions on recovery options, timing of recovery and volume of recovery are taken. So, at the DDP not only the ownership changes hands but also the disposition route is determined.

At first glance one may mistakenly conclude that at the DDP all timing and quantity uncertainty is resolved. However, quality uncertainty may only be resolved after further testing and grading. Sometimes, quality can only be assessed during the recovery process itself. Hence, residual quantity (yield loss: how many units will ultimately be recovered?) and timing uncertainty (how long will it take to recover or decide to dispose?) still remain, even after the physical receipt of the product returns.

Consequently, a larger portion of uncertainty would be resolved at the DDP if more effort was invested there in testing/grading or in collecting information about the state of returns during use (data logging). With regard to the first option, the nature of products or the nature of testing/grading operations may exclude such an option. The returned item may need disassembly in order to assess its quality state, which may be infeasible at the point of receipt. Likewise, testing may require certain equipment or expertise which is only available upstream. On the other hand, data logging and installed base management is mainly possible for electronic products. Furthermore, such approaches do not necessarily result in a definitive conclusion regarding a product's quality state.

To illustrate, consider the recovery of leased copier equipment. Some researchers (e.g., [10, 14]) count leasing among the major success factors for effective implementations of product recovery, because through leasing firms may have a better control of return streams. However, even in the case of lease contracts the control on timing and quantity is not perfect, since the customer may opt to extend the contract. Suppose there is no such option and there are no other sources of returns. Then the firm knows exactly when and how many products will return and plan recovery activities accordingly. The condition of returned items are, however, still unknown and maybe more importantly, non-uniform, even if a lot of items come from the same installation. It may also turn out that some of the cores or components are not recoverable. In the case of component remanufacturing, inspection and testing may require disassembly, which takes place after the core is released to the shop floor. Hence, quality uncertainty carries over the other two types of uncertainties along the CLSC and onto the shop floor.

This persistent variability of the condition of returned items is the main source of complexity in product recovery systems [11]. In the same study, it is noted that, since used products serve as raw materials in such systems, planning and control are much harder than traditional systems, which primarily deal only with demand uncertainty.

In summary, quality uncertainty has a deeper impact on the CLSC and may extend the reach of the other two types of uncertainties. Therefore, as a pre-liminary conclusion, we put forward that research focusing on timing or quantity uncertainty issues in product recovery systems also need to consider quality uncertainty for greater relevancy of the outcomes. This calls for the development of a structured view of quality in product recovery, which is the topic of the rest of this chapter. To do so, we focus on definitions of quality in the next section. First, we present extant quality and (quality based) value definitions from the literature and subsequently relate them to the product recovery environments.

2.3 Definitions of Quality

Garvin [15] provides five definitions of quality, based on different views:

1. Transcendent: This definition refers to "innate excellence". In other words quality is not the result of a specific attribute, but rather a "state".
2. Product based: According to this view quality is a precise and measurable variable. It is a function of certain attributes of the product. The more of those

attributes the higher the quality. The product-based approach is based on performance, features, and durability dimensions of quality.

3. User based: Individuals have different needs and quality is determined by how well these preferences are satisfied. In operations management this view is represented by the "fitness for use" concept. The user-based approach is based on esthetics and perceived quality dimensions as well as more tangible dimensions such as performance and features.

4. Manufacturing based: Unlike the user-based view, which is based on subjective preferences, this view uses objective measures such as tolerances and performance standards. Any deviation from specifications implies a reduction in quality. Conformance and reliability dimensions of quality are the focus of this approach.

5. Value based: In this view quality is jointly determined by the product's conformance to specifications and the price (cost) of attaining that product.

The transcendent view of quality does not provide a tangible scale for measurement. What it lacks is provided by the product-based, user-based, and manufacturing-based views. As noted by Hopp and Spearman [16], these views are product oriented. In other words, they pertain to what is "seen" by the customer. Naturally, it is ultimately important to match what is delivered to what is required. Therefore, a basic tenet is that the product's quality should be defined from the customer's point of view.

Customers have not always been at the focal point of the quality approach. For a long time, the quality concept was rather manufacturing based and the primary concern was detection and control. Due to economic growth, whatever was produced could find a market and only after competition increased customers came into the picture [9]. Delivering what the consumer wants requires more than manufacturing excellence. As opposed to product-based (engineering) and user-based (marketing) definitions of quality that would fit the organizational boundaries, quality is in fact delivered as a result of a process; an aggregate effort from design to manufacturing and then aftersales. This view has led to the emergence of total quality management (TQM).

TQM is a system approach that works backwards and forwards along the supply chain [9]. At the heart of TQM lie the concepts of continuous improvement and "customer value". Customer value extends the value-based definition of quality by incorporating price or costs. Bounds et al. define customer value as "a combination of benefits and sacrifices when a customer uses a service or product to meet certain needs". Those consequences that contribute to meeting one's needs are benefits, while those consequences that detract from meeting one's needs are sacrifices. Thus, Bounds et al. argue that customer value concept encompasses all definitions of quality. To provide value, product designs must conform to customer needs (user-based and product-based quality), manufacturing processes must conform to the designs (manufacturing-based quality), and the product must deliver performance (user-based quality). Value is delivered to the customer during the "use process", which includes all the activities that customers go through in using a product: find, acquire, transport, use, dispose.

The value concept can be extended for product recovery situations. In [13] and [14], "value of returns" is used to define different CLSC strategies. The authors define "negative externality value" and "positive intrinsic value". Negative externality value refers to potential environment or safety risks as well as negative influences on brand image. Positive intrinsic value refers to the "built-in value" and depends on the type of return. In [13] Le Blanc argues that positive intrinsic value is related to the value in reuse for End-of-Use returns and material value for End-of-Life (EOL) returns. Therefore, these definitions are mostly based on what is left of the primary value added process. Thus, from a quality angle they are product-based definitions and not value-based ones.

In [14], a third value definition, namely the time-based value, is used. This notion is based on the time sensitivity concept introduced by Blackburn et al. [17] for commercial returns. It is built on the observation that the value of a product typically diminishes over time. Certain consumer electronics, for instance, may lose 1 % of the original sales value per week. Le Blanc [13] also acknowledges that value can be a function of time. However, unlike Krikke et al. [14], Le Blanc [13] does not present a time-based value definition, but rather argues that value of returns (positive and negative) can be time dependent.

Employing these definitions and using the concepts behind customer value, we can describe "return value" as "a combination of benefits and sacrifices when a product is used in a certain recovery option". In this way the "return value" definition encompasses acquisition of recoverables, reverse logistics activities, and recovery operations which constitute the sacrifices. For example, the existence of a collection network, mandatory take back policies, having a design that facilitates recovery operations, developing a secondary market, etc., define the sacrifices. On the other hand, the intrinsic value of the return, the natural resources saved, improved brand image, etc., define the benefits. The return value definition captures the time dependency of the returns since the recovery options that are available are a function of time. As a summary, a quality-based definition captures the most essential aspects of return value in the recovery processes.

However, the benefits and sacrifices are dependent to a great extent on the recovery option. The recovery options that can be exercised by a supply chain actor are mainly a function of *quality*, a fact that makes the latter the main determinant of return value. Moreover, in order for the value to materialize after recovery, the supply and demand should match. However, an end-user may have a different quality perception of recovered items when compared to newly manufactured items; this gives rise to additional demand uncertainty. Therefore, due to the quality uncertainty of the recoverables the definition of "return value" does not in fact possess the same operational power as "customer value".

This can be explained by the fact that the concept of customer value, which essentially is a "forward" supply chain concept, is well supported by quality definitions and procedures to translate customer needs. Therefore, it is possible to operationalize the concept by segmenting the market and measuring value for each type by marketing tools. Moreover, all the processes in the forward supply chain are tuned to control variation, thus ensuring predictability and stability.

In contrast to this, the "use process" that closes the supply chain is expected to introduce a substantial amount of variability into the chain. Although the acquisition and upstream testing can reduce the variability introduced to the recovery process, the residual variability in turn makes the return value a less useful concept for operational purposes. In other words, incomplete information about quality makes an overarching definition less effective. A similar limitation, likewise stemming from quality variability, was noted by Le Blanc [8] when trying to find out the determinants of the aforementioned DDP. The author concluded that quality information has an obvious relation with where the decoupling point should be placed but the relation is "not unambiguous", thus weakening the DDP concept.

A promising research direction, as pointed out by Krikke [18], is to analyze the "return value" from different value perspectives (see Sect. 2.6). However, since our focus is on the quality aspects of product recovery, we will turn our attention to the building blocks of the value concept in terms of quality definitions. More specifically, in the next section we will focus on the applicability of these definitions to the quality concept along the CLSC.

2.4 Quality Aspects of Product Recovery

Quality, as we have seen, takes on various roles in a CLSC (Fig. 2.1). First, quality problems can be the driver behind the return process itself, such as in warranty returns [5]. Second, the uncertain condition of recoverables introduces variances in material and product flows. Third, the recovery options are defined with respect to certain quality specifications. The perceived quality of the recovered items by the customer/user constitutes the fourth aspect.

For all categories of reverse flows the return reason can be directly related to quality aspects. Warranty returns represent a failure on the reliability dimension of quality. In commercial returns, although the product may be completely functional, there is usually a mismatch with the customer's expectation. This can happen with respect to performance or perceived quality dimensions. For End-of-Use returns the product no longer satisfies the quality expectations of the customer due to changing preferences or better products entering the market. EOL returns are again initiated by failing quality issues due to the product exceeding its technical lifetime. Hence, quality is established as the main driver of reverse streams.

Consequently, the return reason, largely driven by quality aspects, defines the available disposition routes to some extent. For example, if either "esthetics" or "perceived quality" is the lacking dimension as in commercial returns, then the return is possibly routed to direct reuse (that is restocking). If the "features" dimension of quality is lacking the return is a candidate for refurbishing or remanufacturing (e.g., copier remanufacturing). Nevertheless, since the returns are input to the forward flow, a more accurate definition of "product return" is

Fig. 2.1 Depiction of quality
aspects in a generic product
recovery network

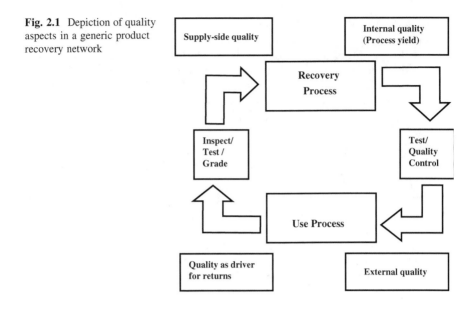

necessary to control the input. Therefore, manufacturing or product-based views of quality with specifications and performance standards are more appropriate. In essence this means that no two items from the same return stream are alike with respect to these specifications, even if they are returned for the same reason.

In Fig. 2.1 we use the term "supply-side quality" for the unknown condition of returned items. The magnitude of uncertainty changes with respect to the type of return stream and the complexity of the product [8]. As Guide points out, as the product gets more complex, simply there are more things that can go wrong. To connect the level of uncertainty and quality as a return reason, we can use the sojourn time of a product with the customer as an explanatory link. Roughly, we can argue that if the sojourn time is relatively short, like with warranty returns, or relatively long, like with EOL returns, the quality and thus the value uncertainty is typically small. In the former, there is a quality complaint on the performance dimension, but the variance is small since the product has not been in extensive use. In the latter, product has no residual functionality or value left apart from material recycling. In between, when the length of the "use process" is more variable, End-of-Use returns pose more quality uncertainty. Without any data logging or product monitoring, it is hard to determine the condition of the product and its components. Moreover, the possibility of damage in transport is higher in the reverse chain since the products are not packaged in a uniform, proper way.

Usually the uncertainty is resolved by checking nominal characteristics and/or inspection-testing efforts. As examples, in copier remanufacturing [19] and cellular phone remanufacturing [20], there are grades defined by nominal characteristics and basic functionality of items. In commercial returns setting, this step is crucial for the decision to refund the customer. In addition, the routing of returns is a result of this

Table 2.1 Definition of product recovery options (adapted from [21])

Recovery option	Quality requirement
Repair	Restore product to working order
Refurbishing	Inspect all critical modules and upgrade to specified quality level
Remanufacturing	Inspect all modules and parts and upgrade to "as good as new" quality
Cannibalization	Recovery of components with "working order" requirements
Recycling	High for production of original parts; less for other parts

activity. In product recovery situations often firms have a number of disposition options available. These options are repair, refurbishing, remanufacturing, cannibalization, and recycling [21]. The best choice among these options is affected by the economics and the incoming quality. Therefore, inspection/testing mechanisms also function as a gateway for the recovery process options.

The third dimension of quality in product recovery systems is internal quality. In Table 2.1, each of the recovery options is defined with respect to the quality requirements to be satisfied. These definitions refer to the conformance to pre-defined standards and therefore are based on the manufacturing-based view of quality. Therefore, once these standards and specifications are in effect, it seems possible to apply the traditional (or primary production) quality processes to recovery situations. However, the nature of the recovery process in most cases dictates performance testing after recovery. In other words, unlike manufacturing (or primary flows) where variations of the processes are stable and controlled, due to the input variability, recovery processes may lead to "defectives" or yield loss. This yield loss may also be the result of the recovery process itself. For example, if the recovery process requires disassembly and the product is not designed to facilitate that operation damaging of the components might occur.

Therefore, after the recovery process, the specified quality requirements are checked by a "test-quality control" step. We will use the term "internal quality" for the process yield issues regarding the recovery operations (Fig. 2.1). It is worth noting that due to the uncertainty from the supply side the test/quality control procedure is essential. In other words, traditional tools such as sampling and control charts are not suitable for the recovery process as they rely on standardized, high quality inputs. Hence, each unit of the output must be checked and tested before meeting demand or reassembly.

The fourth and final quality aspect that we consider in product recovery environments is the "external quality" or the quality observed by the customer. This facet of quality in recovery systems is similar to the traditional usage in manufacturing and engineering aspects. However, there are also additional marketing and strategic issues related to external quality. First, in terms of perceived quality, recovered products or products that have recovered components are sometimes thought of as being inferior. The marketing of these products, with concerns such as market cannibalization and protecting the brand image, may require significant effort. For example, in copier remanufacturing, to convince the customers, remanufactured machines are offered with the same warranty assurances as the new ones. In tyre retreading, this negative perception together with low cost competitors from Far East results in limited use of recovery opportunities for passenger car tyres [22]. Second, if recovery is carried out by independent parties and not done properly it may tarnish the Original Equipment Manufacturer's quality reputation. The case of printer cartridges is an example. Lexmark's move into product recovery and protecting the cartridges by design changes is a result of such concern [23]. Brand protection is one of the main drivers behind car manufacturers' part recovery efforts [24, 25], preventing third parties from recovering their engines without proper quality control.

Summarizing, different definitions apply to the different roles of quality along the CLSC. These are depicted in Fig. 2.2. Returns are triggered by an overall value-based definition of quality. A returned product fails to deliver any more value to the customer. This failure can take place due to any of the dimensions of quality such as performance, reliability, esthetics, etc. Following that, at the input side of the recovery process the product-based definition of quality applies, which relies on performance, features, and durability dimensions. These dimensions, for example features of an electronic product or performance of mechanical parts, are instrumental at the testing/grading step to determine available recovery routes. In other words, the quality of return is measured along certain dimensions. At the output of the recovery process, the internal quality is measured; this time according to manufacturing definitions of quality. There are acceptable tolerances and standards that the recovered product should conform to. When it comes to the external quality or the preferences of the customer, the user-based definition of quality applies; this drives the purchase decision and the return decision. The perceived quality dimension plays a prominent role in this definition and it is more pronounced in product recovery systems than in forward supply chains.

In conventional (forward) production environments quality and quality management have been on the management agenda for quite some time. There is an established philosophy as well as tools and techniques to apply. However, the nature of product recovery environments would not allow the use of some well-known tools like statistical process control in the short run. Given the divergent structure at the "supply side" of product recovery, there is little chance to control incoming variability. This will often lead to unstable variation in the output. Since the recovered components are fed into the primary production lines or completely remanufactured products are delivered to the customers, each item is inspected and tested after the recovery operations. The need to inspect each one of the incoming items (100 % Inspection) and output is reminiscent of early stages of quality control where the focus was on detection. Even with design and process improvements, this is unlikely to change, at least soon. On the other hand, this does not affect the fact that, true to the TQM spirit, the aim and the real benefit should be controlling the variability coming from the previous stages. So at this phase of product recovery where the systems are still developing, the benefit probably lies in utilizing the problem solving and forward-thinking aspects (process thinking, continuous improvement) and organizational aspects (cross-functional management) of quality management. In the next two paragraphs, we will discuss these two aspects.

Fig. 2.2 Quality definitions
along the CLSC

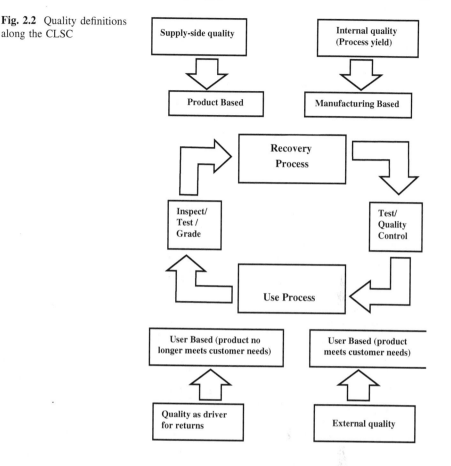

The process-oriented view of quality necessitates the exploration of cause and effect relationships, since quality is a result of an effort along the supply chain. In other words, controlling the quality of inputs to a process is achieved by controlling the previous process. Using this reasoning, what is delivered to the customer ultimately affects the reverse flow. For example, in [26], Klausner and Hendrickson point to the fact that power tools are generally "over-engineered" for an average user, thus recovery of certain components is highly feasible. In the automotive industry, on the other hand, the "built-in obsolescence" concept is used to limit the life of certain components. With European Union regulations holding the manufacturer responsible for EOL vehicles, thus closing the loop, such practices have a direct impact on the feasibility of recovery operations. Therefore, in the product recovery setting, the quality and the value concept should be defined both for the "use process" and the "recovery process". Hence, the product should not only deliver value to the customer, but also should retain a "return value" that can be defined, built-in, measured, and extracted with minimal effort and uncertainty.

From an operational point of view, this closed loop view of quality calls for design improvements by feeding back recovery know-how to designers. In [23], using a resource-based view of the firm, Toffel argues that cross-functional management and continuous improvement capabilities, which are cornerstones of TQM programs, can enable organizations to transfer knowledge that has accumulated in the recovery stage to the design stage. Such approaches will undoubtedly contribute to enhanced quality throughout the CLSC and hence, better decision-making and more return value. In the next section, we turn our attention to managing the product recovery systems themselves.

2.5 Product Recovery Management and Quality Framework

In this section, we relate the quality framework of Sect. 2.4 to the planning and management of recovery systems. This way we establish a link between the quality variability and the planning of recovery systems. To do so, we employ three frameworks from the literature that deal with the planning of product recovery systems [6, 7, 8]. In [8], the characteristics of remanufacturing systems that complicate planning and control activities are discussed. In [7], Geyer and Jackson present their "Supply loop framework", which is used to determine the constraints (or inefficiencies) in the CLSC. Compared to [7] and [8] which address a wider scope of the closed loop supply, Fleischmann et al. [6] focus on the network structures for product recovery and present a classification. In this section, mainly the frameworks in [7, 8] are used to establish a link with our quality framework, while the framework in [6] will be employed in network design-related aspects.

In [8] (and also in [12]), seven major characteristics that significantly complicate management of recoverable manufacturing systems are presented. These characteristics are the following:

1. The uncertain timing and quantity of returns.
2. The need to balance returns with demands.
3. The disassembly of returned products.
4. The uncertainty in materials recovered from returned items.
5. The requirement for a reverse logistics network
6. The complication of material matching restrictions.
7. The problems of stochastic routings for materials for remanufacturing operations and highly variable processing times.

Although these factors are mainly reported for remanufacturing, they fit other types of product recovery as well. These characteristics correspond to constraints on the CLSC in the framework of [7]. A process in the CLSC is constrained if it has difficulties with the output of the upstream process. Therefore, the constraints for the "Acquisition process" (i.e., "access to the products leaving the use process") are characteristics 1, 2, and 5. The constraints for the "Recovery process"

Table 2.2 Complicating characteristics and constraints on recovery networks

Framework Guide [8]	Framework Geyer and Jackson [7]
1. The uncertain timing and quantity of returns 2. The need to balance returns with demands 5. The requirement for a reverse logistics network	Constraint on acquisition
3. The disassembly of returned products 4. The uncertainty in materials recovered from returned items 7. Uncertain routings	Constraint on recovery
2. The need to balance returns with demands 6. The complication of material matching restrictions	Constraint on secondary use

are 3, 4, and 7 and the constraints for the "Secondary use process" are 2 and 6. This mapping is summarized in Table 2.2.

To link our quality framework to the other two frameworks, we discuss the characteristics presented above to find out how they are affected by different aspects of quality, namely "Supply-side quality", "Internal Quality", "External Quality", and "Quality as a driver for returns".

1. *The uncertain timing and quantity of returns.* A return is initiated when the product does not deliver any customer value. Therefore, this characteristic is mainly related to "Quality as a driver for returns". Further, as put forward in Sect. 2.2, the quality uncertainty is the main source of uncertainty for the recovery process since it carries timing and quantity uncertainties over into the process. In this regard, this characteristic is also related to "Supply-side quality".

2. *The need to balance returns with demands.* Firms must balance the customer demand with the returns by either disposing of excess or manufacturing (purchasing) new items to be able to meet demand. Therefore, the variability of "supply-side quality" may necessitate acquisition management activities which in turn are aimed at influencing the flow from the (primary) use process. On the other hand, the lack of an efficient secondary market can constrain the CLSC as noted in the framework of Geyer and Jackson [7]. The "External Quality" aspect of our framework is closely related to the dynamics of the secondary market as explained in the previous section.

3. *The disassembly of returned products.* In [8] this characteristic is discussed as a (re-)manufacturing step before the release of parts to the shop floor only. Fleischmann et al. [6] interpret it in a broader sense to denote all operations for determining whether a given product is reusable and in which way. Using the latter view, the "Supply-side quality" in our framework is related to this characteristic in terms of the level and method of testing required and its location (centralized or decentralized).

4. *The uncertainty in materials recovered from returned items.* This characteristic is directly related to the "Internal Quality" aspect. In other words, it is related to the yield from the recovery.

Table 2.3 Complicating characteristics and Constraints on Recovery Networks

Quality Aspect	Quality Definition	Framework Guide [8]	Framework Geyer and Jackson [7]
Quality as driver for returns	User-Based (product no longer meets customer needs)	1. The uncertain timing and quantity of returns 2. The need to balance returns with demands 5. The requirement for a reverse logistics network	Constraint on acquisition
Supply side	Product based		
Internal quality	Manufacturing based	3. The disassembly of returned products 4. The uncertainty in materials recovered from returned items 7. Stochastic routings	Constraint on recovery
External quality	User-based (product meets customer needs)	2. The need to balance returns with demands	Constraint on secondary use

5. *The requirement for a reverse logistics network.* This characteristic encompasses how products are collected from the end-user and brought to the recovery facility. In their analysis of product recovery networks, Fleischmann et al. [6] note that in traditional production/distribution systems, destination of flows are known with certainty whereas in product recovery networks the routes are dependent on quality and distinguish three types of networks in their analysis: bulk recycling networks, assembly product remanufacturing networks and re-usable item networks. Especially for assembly product remanufacturing networks, the quality of the collected product leads to centralization/decentralization of testing and inspection activities. In this sense this requirement for a reverse logistics network and its design is related to both "Quality as a driver for returns" and "Supply-side quality" aspects.

6. *The complication of material matching restrictions.* This characteristic is specific to the cases where product recovery is serial number-specific. For instance, in aircraft engine recovery certain engine parts need to be recovered as a set. This characteristic does not have a direct relation to quality uncertainty.

7. *The problems of stochastic routings for materials for remanufacturing operations and highly variable processing times.* As Guide [8] points out, variable quality translated as variable processing times make resource planning, shop floor control etc. more difficult. Therefore, this characteristic is related to the "Internal Quality" aspect.

To summarize, out of the seven complicating characteristics reported in [8], six of them are related with the quality uncertainty. In Table 2.3, these relationships are summarized.

Moreover, Table 2.3 establishes a link between definitions of quality and complicating characteristics, and constraints of product recovery networks. As such it gives a comprehensive overview of how the challenges encountered in the CLSC link to the various quality dimensions. Hence, Table 2.3 provides management with powerful handles to actually act on the challenges through quality improvements.

2.6 Conclusions and Future Research

In this chapter, we have analyzed product recovery systems from a quality point of view. We have established quality uncertainty as the main source of variability in product recovery systems. Managing better the quality (uncertainty) in the recovery systems will lead to more efficient CLSC and better use of recovery options. To this end, the framework presented in Sect. 2.4 provides the necessary building blocks and structure by identifying applicable definitions of quality. Furthermore, the link between our quality framework and the existing frameworks in the literature, as established in Sect. 2.5, outline the scope and impact of quality in product recovery systems. Our framework clearly shows that problems and inefficiencies due to quality issues downstream can often be explained through quality management issues upstream in the CLSC. Hence, instead of planning capacities to deal with uncertainties and lack of quality in the recovery process, managers should analyze the processes further upstream to avoid these issues in the first place. The suggested framework can help managers to analyze potential quality bottlenecks in a structured way. Likewise, researchers should investigate the impact of this approach in terms of logistics costs and value creation.

One future line of research can be the in-depth analysis of the impact of quality on the complicating factors as described in Guide's framework [8] and the constraints of Geyer and Jackson's framework [7]. This could be done along different dimensions such as different types of returns (EOL, End-of-Use, commercial, and warranty returns) and different product types (electronics, automotive etc.). Another important line of research is to analyze the "return value" from different perspectives, for instance as coined by Krikke [18] (sourcing value, environmental value, customer value, and informational value), to quantify the impact of quality and quality uncertainty on value creation. For this, however, a proper framework needs to be developed, defining the value dimensions and assigning proper constructs to measure value along these dimensions.

Acknowledgments We thank the Dutch Institute for Advanced Logistics (DINALOG; www.dinalog.nl) for their support of this research.

References

1. Guide VDR Jr, Van Wassenhove LN (2009) The evolution of closed-loop supply chain research. Oper Res 57(1):10–18
2. Kumar V, Shirodkar PS, Camelio JA, Sutherland JW (2010) Value flow characterization during product lifecycle to assist in recovery decisions. Int J Prod Res 45(18–19):4555–4572
3. Visich JK, Li Suhong, Khumawala BM (2007) Enhancing product recovery value in closed-loop supply chains with RFID. J Manag Issues 19(3):436–452
4. Zuidwijk R, Krikke HR (2008) Strategic response to EEE returns: product eco-design or new recovery processes? Eur J Oper Res 191(3):1206–1222
5. De Brito M (2003) Managing reverse logistics or reverse logistics management? ERIM Ph.D. series research in management, vol 35. Erasmus research institute of management, Erasmus University Rotterdam, Rotterdam
6. Fleischmann M, Krikke HR, Dekker R, Flapper SDP (2000) A characterization of logistics network for product recovery. Omega 28(6):653–666
7. Geyer R, Jackson T (2004) Supply loops and their constraints: the industrial ecology of recycling and reuse. Calif Manag Rev 46(2):55–73
8. Guide VDR (2000) Production planning and control for remanufacturing. J Operat Manag 18:467–483
9. Bounds G, Yorks L, Adams M, Ranney G (1994) Beyond total quality management: toward the emerging paradigm. Irwin/McGraw-Hill, New York
10. Thierry MC (1995) An analysis of the impact of product recovery management on manufacturing companies. Ph.D. series in general management, vol 26, Rotterdam school of management, Erasmus University Rotterdam, Rotterdam

11. Guide VDR, Van Wassenhove LN (2001) Managing product returns for manufacturing. Prod Operat Manag 10:142–155
12. Guide VDR, Jayaraman V (2000) Product acquisition management: current industry practice and a proposed framework. Int J Prod Res 38(10):3779–3800
13. Le Blanc HM (2006) Closing loops in supply chain management: designing reverse supply chains for end-of-life vehicles, Ph.D. Thesis. Tilburg University, Tilburg
14. Krikke HR, Le Blanc HM Van de Velde SL (2004) Product modularity and the design of closed loop supply chains. Calif Manag Rev 46(2):23–39
15. Garvin D (1988) Managing quality: the strategic and competitive edge. Free Press, New York
16. Hopp WJ, Spearman ML (2000) Factory physics. Irwin/McGraw-Hill, New York
17. Blackburn JD, Guide VDR, Souza GC, Van Wassenhove LN (2004) Reverse supply chains for commercial returns. Calif Manag Rev 46(2):6–22
18. Krikke H (20090 Opposites attract: a happy marriage of forward and reverse supply chain. Inaugural speech, Open University, Heerlen
19. Krikke HR, van Harten A, Schuur PC (1999) Business case oce: reverse logistic network redesign for Copiers. OR Spektrum 21:381–409
20. Guide VDR, Van Wassenhove LN (2003) Business aspects of closed loop supply chains. In: Guide VDR, Van Wassenhove LN (eds) Business aspects of closed loop supply chains. Carnegie Mellon University Press, Pittsburgh, pp 17–43
21. Thierry MC, Salomon M, van Nunen JAEE, van Wassenhove LN (1995) Strategic issues in product recovery management. Calif Manag Rev 37(2):114–135
22. Debo LG, Van Wassenhove LN (2005) Tyre recovery: the retread Co case. In: Flapper SDP, van Nunen JAEE, van Wassenhove LN (eds) Managing closed-loop supply chains. Springer, Berlin, pp 119–129
23. Toffel MW (2004) Strategic management of product recovery. Calif Manag Rev 46(2):120–141
24. Driesch H, van Oyen HE, Flapper SDP (2005) Recovery of car engines: the Mercedes-Benz case. In: Flapper SDP, van Nunen JAEE, Van Wassenhove LN (eds) Managing closed-loop supply chains. Springer, Berlin, pp 157–169
25. Seitz MA (2007) A critical assessment of motives for product recovery: the case of engine remanufacturing. J Clean Product 15(11–12):1147–1157
26. Klausner M, Hendrickson CT (2000) Reverse-logistics strategy for product take back. Interfaces 30(3):156–165

Chapter 3
Standardization of the Reverse Logistics Process: Characteristics and Added Value

Ir André Pirlet

Abstract Reverse Logistics initiatives aim at solving, at least partially, important problems linked to economical savings, durability, and environmental concerns. To improve the reverse logistics processes, a comprehensive approach is frequently needed, including standardization. Standards contain unambiguous requirements, and also, when needed, detailed testing methods. The European Standardization Organizations (ESO) offer a choice of ways for efficiently reaching consensus, in the form of high status written documents called European Standards, Technical Specifications or CEN Workshop Agreements. It is best to reflect early on the need for amended or new standards, and, if such a need exist, to start the standardization process in many cases as soon as possible. This applies also to EU research and technical development projects, which can usefully encompass a Work Package Standardization. Standardization can represent a key aspect of an improvement project in reverse logistics, in particular when harmonized requirements and/or testing methods are really needed. So far there seems to be no existing large-scale standards in that field, at least at the European or worldwide level. This chapter begins with a summary of standardization characteristics and advice for choosing the best procedure. It continues with a description of challenges to tackle in reverse logistics, and a review of existing standards and standards structures in that field. This is followed by a methodological approach to improve the current situation, relying in particular on the added value of the so-called "Integrated Approach", to get large-scale beneficial impacts.

Ir A. Pirlet (✉)
CEN-CENELEC Management Centre, Brussels, Belgium
e-mail: a.pirlet@hotmail.com
URL: www.cen.eu; www.cenelec.eu

Y. Nikolaidis (ed.), *Quality Management in Reverse Logistics*,
DOI: 10.1007/978-1-4471-4537-0_3, © Springer-Verlag London 2013

3.1 Introduction

Reverse Logistics (RL) is frequently defined as the coordination and control, physical pickup and delivery of the material, parts and products from the field to processing and recycling or disposition, and subsequent returns back to the field where appropriate. RL initiatives aim at solving, at least partially, important problems, linked to economical savings, durability, and environmental concerns.

In the context of progressive depletion of physical resources (oil, iron, uranium, potable water, etc.), and also, to a lesser extent, climate change, it makes sense to minimize the net use of these resources through the best approach and best management of RL.

In this chapter devoted to the standardization contribution for reverse logistics, Sect. 3.2 first explains the main characteristics of standardization, in order to specify the existing standardization bodies and the alternative standardization deliverables to identify the best procedures for specific objectives, then Sect. 3.3 analyzes some important challenges in RL. In Sect. 3.4, we present the possible structures for developing new standards, and finally in Sect. 3.5 we present a framework which can improve the efficiency in the RL management. Rather than being of a purely scientific nature, which it is not, this chapter can serve to enhance the use of best practice guidelines and therefore contribute to a beneficial long-lasting impact.

3.2 Standardization: The Main Characteristics and the Choice of the best Procedure

Standards are defined as documents established by consensus that provide, for common and repeated use, rules, guidelines, or characteristics for activities or their results, aimed at the achievement of the optimum degree of order in a given context. Standards should be based on consolidated results of science, technology, and experience, and aimed at the promotion of optimum community benefits. Standards contain unambiguous requirements, and also, when needed, definitions and testing methods, to assess whether the prescribed requirements are fulfilled. As they are built on consensus, groups of experts need to meet and communicate. This calls for management, rules, and procedures, which form a standardization framework. Non-formal standards can be written by consortia, whether national, European, or worldwide.

Formal Standards can be defined as standards prepared using formal procedures of openness and transparency and published by permanent and for non-profit recognized standardization organizations. Most of the countries worldwide have a National Standards Body. At European level three European Standardization Bodies can be found:

- CENELEC for Electro-technical Standardization,
- ETSI for Telecommunications, and
- CEN for the rest.

At worldwide level, we similarly have

- IEC for Electro-technical Standardization,
- ITU for Telecommunications, and
- ISO for the rest.

It is essential for standards to be written in such a way they allow evolution and progress, and do not block innovation. There should be flexibility for meeting the requirements specified in the standards; therefore, the modern emphasis on "performance standards". Similar possibilities are offered by CEN and CENELEC, while the ETSI standardization system is rather different and focuses only on telecommunications.

An EN is a European Standard, and enjoys the highest status. A very important aspect is that when an EN is adopted by CEN or CENELEC, their Members who are National Standardization Bodies are forced to adopt the full EN as one of their national standards, and also to withdraw any conflicting national standard. An EN is issued by CEN-CENELEC in three official languages (French, English, and German), but the National Standardization Bodies in Europe generally issue these standards in their national language(s), and this is a key advantage which is frequently overlooked. A Technical Specification (TS) is a prospective standard for provisional application. It is mainly used in fields where the innovation rate is high, or when there is an urgent need for guidance, and primarily where aspects of safety for persons and goods are not involved. Conflicting national standards may be kept in force in parallel with the national implementation of the TS. A Technical Report is a non-normative CEN (or CENELEC) publication authorized by the CEN (or CENELEC) Technical Board. A CEN Workshop Agreement (CWA) is a document prepared rapidly by experts, without formal consultations (Enquiry, Formal Vote) at national level. It is a frequent standardization deliverable from research projects. This type of publication aims at satisfying market demands for a more flexible and timelier alternative to the traditional EN, but it still possesses the authority derived from the openness of participation and agreement inherent in the operations of CEN. These CWAs are produced in flexible structures called CEN Workshops, where the registered participants are in charge of both the drafting and the management. CWAs are particularly suited for the exploitation of results of Research Projects, and that approach is much appreciated by research consortia.

Formal standardization takes place in Technical Committees (TCs), where national delegations are in charge of the management, while the drafting of standards is made by experts sitting in Working Groups, reporting to their relevant TC. The cross-fertilization nature of standardization committees, due to the involvement of researchers and the various stakeholders, is felt as an additional benefit. Whenever possible, preference should be given to the drafting of performance standards, which are defined as standards where requirements allow evolution and progress, and do not block but rather enhance innovation. They offer flexibility for meeting the requirements.

The aforementioned deliverables are quite similar to those found in the International Standardization Organizations ISO and IEC. The most important difference is that there is no mandatory national implementation of the standards published by ISO and IEC. When wishing to initiate a new standardization activity, it is important to take into account the pros and cons of these deliverables, and subsequently choose the most relevant procedure in a tailored manner.

After having explained the basics of standardization, attention can now be turned to the challenges of RL.

3.3 Challenges to Tackle in Reverse Logistics

RL faces various main challenges, noting that the ultimate global challenge is to satisfy company goals while at the same time reaching important societal benefits. At company level, a distinction can be made between "intrinsic" motivation, and "extrinsic" motivation [1, 2]. RL is part of an efficient Supply Chain Management, which is defined as management of a network of interconnected businesses involved in the ultimate provision of product and service packages required by end customers.

The first challenge is for a company to enhance its competitiveness. Through better design and management of its reverse logistics processes, a company should reach substantial savings. The field here is very wide, but let us just give a typical example. Retailers selling perishable food products need to make an accurate guessing of their future sales, in order to narrow the gap between actual offer and actual demand. Avoiding to have outdated products minimizes the problem of dealing with the leftovers, and maximizes the profit margin. Moreover, consider the problem of correct disposal of out-dated medication.

The second challenge is climate change. Accurate measurements show that there is a continuous increase in the quantity of CO_2 in the air. Notwithstanding the scientific controversy on the exact effects that CO_2 increase has on temperatures worldwide, a CO_2 increase in the air has potential negative consequences in the form of increased acidification of the air and of the oceans! Improving recyclability will help to minimize the energy used for manufacturing, including fossil energy, and therefore have a positive impact.

There is not only an overall concern to use less energy, but also, more generally, to use less raw materials, not just to save costs, but also because there is a finite amount on this planet, and the more we extract those raw materials, the more difficult and costly it will be to get new ones. Here again more efficiency in recyclability will help. This is addressed by the concept of sustainable Supply Chain Management [2].

Similar considerations do apply when we consider various types of environmental pollution such as air, soil, or even just visual pollution (e.g., the problem caused by inadequate handling of refuse in the Naples region). A real potential exists for substantial decrease of any type of pollution through better management of reverse logistics, creating for example the best conditions to give a second life to valuable products. In that respect, the industry of recycled paper is one of the most prominent examples of success, in particular since this was achieved not just through industrial investments and commitments but also through the support of consumers. To show how wide and varied are RL, let us mention the example of the growing problem caused by debris from deficient or end of life satellites, with damaging consequences for other satellites, or the possibilities of batteries exchange to help to develop the market of electrical vehicles!

For each type of product return, Guide and Van Wassenhove [3] explain that there is a most attractive recovery option, bearing in mind that recovery actions include repair, refurbishing, remanufacturing, cannibalization, and recycling. Commercial returns have barely been used and are best reintroduced to the market as quickly as possible. The majority of these returns require only light repair operations (cleaning and cosmetic). End-of-use returns may have been used intensively over a period of time and may therefore require more extensive remanufacturing activities. The high variability in the use of these products may also result in very different product disposition and remanufacturing requirements. Ideally, one would like to acquire end-of-use products of sufficient quality to enable profitable remanufacturing. End-of-life products are however predominantly technologically obsolete and often worn out. This makes parts recovery and recycling the only practical alternatives (assuming that one wants to avoid landfill). Summarizing, there are natural return-recovery pairs: consumer returns → repair, end-of-use returns → remanufacture, and end-of-life returns → recycle.

Closed loop supply chains (CLSC) focus on taking back products from customers and recovering added value by reusing the entire product, and/or some of its modules, components, and parts. Over the past 15 years, CLSC have gained considerable attention in industry and academia. Today, CLSC management is defined by Guide and Van Wassenhove [3] as the design, control, and operation of a system to maximize value creation over the entire life cycle of a product with dynamic recovery of value from different types and volumes of returns over time. CLSC activities are examined in detail in the authoritative and most recent book of Ferguson and Souza [4], which investigates the state of the art in this rapidly growing and significant environmental field. Remanufacturing, recycling, dismantling for spare parts, and RL have helped indeed many companies tap into new revenue streams by finding secondary markets for their products, all while reducing their overall carbon footprint. More specifically, remanufacturing is defined as the process of disassembly and recovery at the module level and, eventually, at the component level. It requires the repair or replacement of worn out or obsolete components and modules. Products can also be resold in a different form but, very often, it is not technical constraints that matter, but rather the lack of a market for remanufactured products or the lack of used products of sufficient quality at the right price and the right time. Gatekeeping is defined as the screening of products entering the RL pipeline. When companies need to manage returns across international borders, reverse logistics is a more complex problem. The complexity and the cost of freight can outweigh the benefits of receiving the returns, and preference is then given to a zero return policy, meaning that the manufacturer never takes possession of returns. These "returns" are sometimes destroyed in the field by retailer or third party. At this point, consider also the Centralized Return Centres (CRCs), which form a perfect example of smart cooperation. The trend for separate forward and reverse distribution centres is noted because reverse CRCs are more profitable for the company when outsourced. When both functions occur in the same center, forward supply takes precedence over returns. Typical benefits from a reverse CRC include:

- simplified storage procedures
- improved supplier relationships
- better returns inventory control
- improved inventory turns
- reduced administrative costs
- reduced store level costs
- reduced shrinkage, refocus on retailer core competencies
- visibility of quality problems
- reduced landfill and improved management information.

The effectiveness of third-party logistics-run CRCs results from the economies of scale from serving multiple companies. Specific standards can help to avoid problems and maintain sustained trust, reliability, and efficiency. As it has been already seen above, firms can also develop "zero return" programs, to avoid physically accepting the returns. Returns can then show up in alternative channels.

Another problem that should not be ignored is the product recall, which is defined as the act of requesting the return of a commercial product to the retailer or manufacturer because of a defect or a safety or an efficiency problem. Companies need to have in place expert solutions for promptly and effectively managing the entire product recall process, with limited loss of reputation and profits. This implies a customized, best practice-based recall protocol, and also developing customized notification packages for all recipients, managing database integration and industry mailing list, overseeing rapid distribution of direct mail notifications, managing response processing and reporting, managing notification efficacy check/reporting, facilitating all reimbursements, and disposing/handling of products. These problems seem to concern in particular the automobile sector and the sectors of food and medications. For examples of such existing formal procedures for mandatory recalls, one can refer to Regulation EC 178/2002 concerning food safety as well as to some other guidance issued by the European Food Safety Authority (EFSA), by the MONIQA Network of Excellence [5], or other experts [6], or to standards and procedures used by national product recall agencies [7].

Companies will not just save money and troubles, but also enjoy a better "image", if they convince their customers and potential customers that they duly consider all the concerns and steps mentioned above. Indeed, although RL has existed as long as forward logistics, growing social interest for the environment has caused reverse logistics activities to become a critical function for many organizations. Finally, it seems not farfetched to imagine these companies could get some rebate for their insurance premiums, since such moves would reduce certain risks.

For governments, the challenge is, among others, to create or adapt their legislation, to minimize the problems linked to reverse logistics. European legislation can also help—consider for example the ELV European Directive—dealing with the end-of-life vehicles [8], or the Directives on Packaging and on Dangerous Goods, or the WEEE one, for Waste Electric and Electronic Equipment. Each Member State in the European Union needs to adapt that form of legislation (i.e. the European Directives) at the national level.

This section has provided a summary view of known main RL challenges, and it is up to each group of stakeholders, and to nobody else, to put priorities and identify, in more detail, their needs and how to address them, using the methodology described in detail in Sect. 3.5. First, Sect. 3.4 will set the scene on the standardization structures and on existing standards for RL.

3.4 Existing Standardization Structures and Standards in Reverse Logistics

Yet, there exists no important wide scope, European, or worldwide standards, devoted specifically to the field of RL. Some initiatives have been taken most probably at national level in certain countries. Let us mention in particular the work which has been launched a few years ago in the US, by the Reverse Logistics Association, which did set up a Standards Committee. That Standards Committee deals particularly with terminology and definitions, with return guidelines and best practices. Their work will help the development of a reverse logistics process and cost model, agreeing on a return authorization format. It will also help to collect comparable data and draft return guidelines, as well as guidelines for reconciliation between retailers and manufacturers. But there is also a European group, Rev Log, which is a cooperation of some Dutch, Greek, German, and French Universities: Erasmus University Rotterdam, Aristotle University of Thessaloniki, Eindhoven University of Technology, INSEAD Fontainebleau, University of Magdeburg, and Piraeus University. The aim of Rev Log was to develop a framework for research in the field of RL, which studies the effective integration of product return flows in the production cycle. Research in this field is very much multidisciplinary and incorporates, among others, managerial, environmental, legislative, and logistical aspects. It considers a wide spectrum of products, ranging from multi-million machines to cheap packaging materials. Setting up a management research field—which is of major importance for our society today—requires a coherent classification of relevant research subjects and their interrelations.

Though not specifically written to improve RL, some other, formal standards will nevertheless benefit that field. Let us mention for example the work done within the Technical Committee CEN/TC 261 on "Packaging", which has produced standards for packaging material recycling or reuse, or the detection of heavy metals in packaging, or on material recovery and life cycle inventory analysis. Of course, we should also mention the Technical Committees

- CEN/TC 319 "Maintenance",
- CEN/TC 320 "Transport—Logistics and Services" and
- CEN/TC 379 "Supply Chain Management",

which could form the most logical structures to help improve that lack of existing standards for reverse logistics.

CEN/TC 319 has gathered experience on the competencies of personnel active in maintenance, and it will soon start the drafting of a standard for the three necessary qualification levels, namely the European Maintenance Technician (European Quality Foundation level 5), the European Maintenance Supervisor, and the European Maintenance Manager, all these three levels being formally recognized by the European Union. CEN/TC 320 has a very wide scope and deals, among others, with the application of quality assurance standards to the transport and distribution industries. CEN/TC 379 focuses more on security issues in supply chain management, contributing therefore to the Mandate addressed to the three ESOs by the European Commission, with the support of the Member States, for developing missing standards in "security".

Choosing the best sampling procedure is always a challenge. Here, the series formed by the ISO 2859 standards help the manufacturer to assess whether consumers will find certain products acceptable. In some settings, the cost of examining each item on an assembly line would be unreasonable in terms of time and cost. ISO 2859 [9] discusses how to sample items off the production line and estimate whether the customer will accept a specific batch of goods, without examining every item. The methods presented in ISO 2859 focus on the product's attributes, which represent characteristics. An inspector samples a few items from a lot, studies the condition of an attribute, and computes the probability that the customer will accept the entire lot. The decision relies on statistical analysis. ISO 2859 defines different sampling approaches and indicates which computations to run based on how many defective items surfaced. Then, on the basis of the number of defects observed, ISO 2859 indicates whether the inspection can be relaxed or tightened. In certain cases, it is worth enhancing the quality of sampling through the use of reference materials [10]. ISO 2859:1999 was criticized by von Collani on a variety of reasons [11], but its part 1 was nevertheless reconfirmed by ISO in 2009. An additional useful series is given by ISO 3951 "Sampling procedures for inspection by variables" [12]. These series, namely ISO 2859 and ISO 3951, can qualify as "horizontal" standards for quality control, since they apply to very different products. Moreover, they are valid both for forward and reverse logistics. In most cases, however, the heterogeneity of products and their conditions is much higher in RL than in forward logistics, and, therefore, this impacts not only the sampling modalities to select, but also influences the drafting of additional requirements and testing methods, in standards, to provide the necessary level of guarantee of quality. This is precisely why there is a need for each relevant stakeholder group to systematically analyze the standards produced originally for classical forward logistics, case by case in relation to their objectives, and decide either to suitably add in these standards specific requirements and testing methods to cover reverse logistics, or to draft specific new standards focused on RL and CLSC.

In addition, a large benefit is provided by more general ISO quality standards, such as ISO 9000 [13], ISO 14000 (Quality management for the environment), [14] and ISO 26000 (Social responsibility) [15], since the concerns for the environment [16] and for corporate social responsibility become ever more important for the authorities and the public at large.

In Sect. 3.2, we have briefly described the added value of the CEN Workshop Agreements, i.e., the CWAs. At least in a first stage, this is also a convenient approach to use in RL, to start the harmonization.

3.5 Improving the Current Situation through Project Approaches and Implementation

Large-scale beneficial changes rarely happen by chance, but result from a set of well-planned moves and actions. The "driving" forces should logically be the groups penalized by the present situation and who could expect substantial improvements by taking the right steps. There is therefore a need to use a *Project Approach*.

The starting point should therefore clearly define the difficulty and the challenge(s) to address in RL. Then, there is a need for that group to express a clear objective, which should logically be to minimize or to entirely suppress the difficulty. Once there is a consensus within the group on the objective to be reached, the necessary time must be allocated to the careful drafting of the corresponding "business plan". To reach an ambitious objective, a multi-faceted approach should frequently be used. Some results of research might be needed, whether from already carried research, or from current research projects or already planned research. In certain cases it will be necessary to initiate specific new research activities. Then the issue of legislation comes. A current legislation might apply and the stakeholder group should examine how to adapt to any new legislation, and also whether amendments are needed (or are desirable) for reaching their objective, or no legislation yet exists but is nevertheless needed and should be put in place, for example a European wide legislation.

The same approach applies to standardization. Amendment of existing standards or drafting of entirely new standards might be a necessity. This would take place rarely at a purely national level, but more and more frequently at the European or worldwide level.

Finally, but quite importantly, the "complementary measures" come. Here, we find actions such as marketing, education, training, promotion, protection of

Fig. 3.1 The integrated
approach

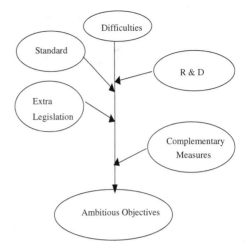

"Intellectual Property Rights", etc., in order to have a comprehensive approach, also called as Integrated Approach.

Using such "Integrated Approach" will give the confidence that the goals can be reached in practice, in an efficient way. Standards can provide clarity on issues of terminology and definitions, durability, recyclability, sharing and comparing data, warranties, counterfeiting, and enable interoperability and economies of scale. They can also form the basis for rules of best practice. Standardized protocols and improved information technology material compatibility will also do wonders to bring more progress and efficiency in reverse logistics, as explained by Pirlet [17]. Some benefit should also be expected from more sophisticated databases and from the progress of the semantic Web. The methodology of "Integrated Approach" can be summarized through the diagram shown in Fig. 3.1

Of course, having the right business plan, using this Integrated Approach concept, is a good start, but the "driving force" needs to select the right people to ensure the best implementation of the plan, within the decided time frame. This calls also for careful monitoring of the implementation, possibly some amendments and contingency measures, and in some cases the use of the AHP methodology [18], i.e., the Analytic Hierarchy Process. A real-world example of using the Integrated Approach just described in a domain involving RL, is given by the battery switch changes, for electrical vehicles. This requires a complex organization, it has necessitated lots of research for improving the range, durability, and reasonable cost of these batteries, some standardization at the national level (e.g., Israel) before Worldwide standardization through IEC, some complementary measures in terms of marketing and training of personnel, and it is subjected to various legislation (LVD, EMC, etc.). Tyre recycling is another interesting field, where reverse logistics has obviously a fundamental place. A proactive role has been played by the European Tyre Recycling Association (ETRA), which did not just lobby actively on the legislation side but did also address the need for standard, first through a CEN Workshop Agreement, and later through a Technical

Specification, namely CEN/TS 14243:2010. TS 14243 [19] specifies the standard materials which can be produced from end of life tyres.

At this stage, it is good to briefly address the "motivation" factor. In research activities, people are frequently "super motivated", even passionate, this "comes with the territory". Motivating legal experts to work on new pieces of legislation is in general not too difficult. On the other hand, it is important to use a careful message in order to ensure a long-term commitment in a standardization activity, frequently characterized as "non-sexy", almost jokingly! People need first to be convinced by the intrinsic importance of the end goals, and second need to be convinced that a "support through standards" is vital to help reach these goals. The importance of these goals will then generate the necessary motivation. It is however harder and harder to find the right kind of people for writing new standards: they must combine solid technical experience with the necessary patience and tenacity, while also being at ease with wordings to formulate precise requirements and explain in all necessary details the most relevant testing methods. To compound the difficulties, these persons should be generally multi-skilled, and might be "derouted" from their standardization tasks to fulfill instead a set of shorter term goals, as decided by their hierarchy.

The larger reliance on outsourcing and subcontracting in present manufacturing industries complicates in a way the drafting of standards, but reinforces on the other hand the benefit which can be reached through these new standards. The CORELOG project [20] has aimed at producing transnational strategic guidelines for regional policymaking and implementation of sustainable freight transport through public–private co-ordination, and also related checklists for policymaking in transport and logistics, considering actors and needs, impacts and benefits on territories, and critical success factors and partners' experiences. That project has shown how cooperation among companies, in terms of vertical integration of the supply chain activities and horizontal cooperation among companies in specific clusters and industrial areas, can bring higher profits and environmental gains in terms of reduction of transport externalities. These guidelines from the large CORELOG project already encompass compromise views due to the input of the various members of the consortium and therefore constitute useful documents for transformation into formal standards, when needed. It has been stressed in [20] that in logistics, training, education, and the certification of skills and competences must be provided through a well-established framework program from educational entities, professional associations and relevant ministries. This also applies to RL, which should be part of that training. Measures aiming at the promotion of harmonized classifications for logistics jobs will improve the overall transport and logistics environment and the quality of services, giving a more structured establishment of logistics jobs, covering also reverse logistics.

Finally, in terms of spent resources, we should stress that, compared to some large expenses such as research or marketing, standardization is relatively cheap. The standardization costs could then be easily envisaged, even in the absence of financial support (from the European Commission or other third party).

3.6 Conclusions

This chapter has been an introduction to the main characteristics of standardization and its potential benefits for the field of reverse logistics. Standardization is indeed frequently overlooked, in spite of its large impact. We have shown the need for quality procedures (and therefore standards), to ensure smooth and economical operation in reverse logistics. A message has been passed here that main stakeholders in the "real world" should feel encouraged not only to examine how to use existing standards, but also to consider the needs for new standards and to be ready to initiate their drafting. This is feasible, even if this is rarely straightforward. Second, we hope to have been convincing on the interest to use an "Integrated Approach" to reach ambitious complex objectives, with much greater chances of success, be it in the field of reverse logistics, or for that matter in many other complex applications.

References

1. RLEC, Reverse logistics executive council glossary http://www.rlec.org/glossary
2. Teuteberg F, Wittstruck D (2010) A systematic review of sustainable supply chain management research: what is there and what is missing. In: MKWI 2010—Betriebliches Umwelt und Nachhaltigkeit management, pp 1001–1015
3. Guide VD, Van Wassenhove L (2009) The evolution of closed-loop supply chain research. Oper Res 57(1):10–18
4. Ferguson M, Souza G (2010) «Closed-loop supply chains: new developments to improve the sustainability of business practices. XVIII in Supply chain integration: modeling, optimization and applications, CRC Press, Boca Raton pp 1–239 ISBN 978-1-420-09525-8
5. www.moniqa.org
6. Lineback D, Pirlet A, Van der Kamp J-W, Wood R (2009) Globalization, food safety issues and role of international standards. Qual Assur Safety Crops & Foods 1(1):23–27
7. Drake M, Mawhinney J (2008) Reverse logistics strategies in the United States. In: Proceedings of the international conference on reverse logistics and global closed-loop supply chains, Beijing, P.R.C., pp 180–190
8. Krikke H (2010) Opportunistic versus life-cycle-oriented decision making in multi-loop recovery: an eco–eco study on disposed vehicles. Int J Life Cycle Assess 15:757–768
9. ISO 2859 series (2009) Sampling procedures for inspection by attributes. www.iso.org
10. Emons H (2005) Quality control of sampling using reference materials. In: Chemical and environmental sampling-quality through accreditation, certification and industrial standard, CEN/STAR trends analysis workshop in co-operation with Nordic Innovation Centre, Brussels, April 14–15 2005
11. von Collani E (2004) Review of statistical standards and specifications-A problem called ISO 2859-1 Sampling procedures for inspection by attributes Part 1. Econ Qual Control 19(2):265–276 ISSN 0940-5151
12. ISO 3951 series (2006) Sampling procedures for inspection by variables. www.iso.org
13. ISO 9000 series (2008) Quality management. www.iso.org
14. ISO 14000 series (2009) Environmental management. www.iso.org
15. ISO 26000 (2010) Guidance on social responsibility. www.iso.org
16. Ilgin MA, Gupta SM (2010) Environmentally conscious manufacturing and product recovery (ECMPRO): a review of the state of the art. J Environ Manage 91(3):563–591

17. Pirlet A (2009) Standardization as an ICT implementation enabler. In: 2nd European conference on ICT for transport logistics, 29–30 Oct 2009, San Servolo Venice
18. Fernandez I, Gomez A (2005) Analisis empirica de la logistica inversa. Una applicacion de la metodologia AHP. IX Congreso de Ingenieria, Gijon, 8–9 sept 2005, pp 1–15
19. TS 14243 (2010) Materials produced from end of life tyres. Specification of categories based on their dimension(s) and impurities and methods for determining their dimension(s) and impurities. www.cen.eu
20. Measures and Actions for Coordinated Regional Logistics Policies In: The framework of the CORELOG EU Project Interreg III B CADSES NP, edited by Direzione Generale Reti Infrastrutturali, Logistica e Sistemi di Mobilità, Regione Emilia-Romagna (in cooperation with all CORELOG partners)

Chapter 4
A Framework for Evaluating the Social Responsibility Quality of Reverse Logistics

Ioannis E. Nikolaou and Konstantinos I. Evangelinos

Abstract This chapter proposes the combination of basic principles of corporate social responsibility (CSR) with reverse logistics (RL) systems as a means of developing a social responsibility quality framework for evaluating RL procedures. More specifically, this chapter provides an overall framework including indicators for measuring reverse logistics social responsibility quality based on the triple bottom line approach (namely economic, environmental, and social aspects). This framework helps to tackle the weakness of the present environmental and sustainable RL systems. These are mainly based on measuring a limited number of social aspects, which are necessary for having a clearer picture of the quality performance of RL systems. The suggested framework could also play a critical role in measuring the contribution of a firm to sustainable development quality within the RL context.

4.1 Introduction

In the last decade, the business community and academics have become increasingly interested in reverse logistics (RL) systems. Ravi and Shankar [1] state that the growing interest in RL is associated with competition, marketing, economic,

I. E. Nikolaou (✉)
Department of Environmental Engineering, Democritus University of Thrace,
Vas Sofias 12, 67100 Xanthi, Greece
e-mail: inikol@env.duth.gr

K. I. Evangelinos
Department of Environmental Studies, University of the Aegean, University Hill,
Mytilini, Greece
e-mail: kevag@aegean.gr

Y. Nikolaidis (ed.), *Quality Management in Reverse Logistics*,
DOI: 10.1007/978-1-4471-4537-0_4, © Springer-Verlag London 2013

and environmental factors. In this logic, many manufacturers adopt RL management practices aiming to improve both the financial and environmental performance of their firms so as to reduce costs, improve their firms' profile, yield a competitive advantage, have fewer environmental impacts, and improve their environmental quality of production. Legislative factors also explain businesses' decision to adopt RL practices. Gonzalez-Torre et al. [2] note that the European Directive 94/62/EC on packaging and packaging waste has forced manufacturers to recover a significant amount of packaging and waste of the products they market. Subramanian et al. [3] claim that electronic and electrical manufacturing face the dilemma whether to comply with the European Directive 2003/108/EC for managing electrical and electronic waste as well as with 2002/96/EC for the restriction of hazardous substances. To this end, Gottberg et al. [4] identify that the European Directive 2002/96/EC is a good driving force for the implementation of eco-design by high-polluting businesses.

Some researchers have studied RL in association with the concept of environmental management quality and corporate social responsibility (CSR) quality practices. Tsoulfas et al. [5] conducted an environmental analysis of the battery sector, using life cycle analysis in order to improve the recovery process within RL. Georgiadis and Vlachos [6] propose a tool based on system dynamics to study how RL processes could be affected by external factors, such as environmental legislation and investments in remanufacturing facilities. Presley et al. [7] suggest the introduction of issues relevant to the three major sustainability pillars, namely financial, environmental, and social, in an RL decision-making process. However, RL systems have been examined in relation to CSR practices and a series of academic papers show how CSR principles can be introduced in logistics and supply chain of firms. For instance, Ciliberti et al. [8] classify the logistics social responsibility practices of firms in five categories: Purchasing social responsibility, sustainable transportation, sustainable packaging, sustainable warehousing, and RL systems. Similarly, Malony and Brown [9] developed a comprehensive framework of supply chain CSR in the food industry.

However, those models have some shortcomings. For example, the majority appears to deal only with environmental and economic parameters, excluding the social ones [10, 11]. Presley et al. [7] state that some of the present research

> focuses on strategic proactive measures incorporating life cycle analysis, and others on end-of-pipe, traditional waste management technology evaluations. None of those tools, however, explicitly introduce additional sustainability concerns such as social sustainability into their analyses

(p. 4598). Additionally, the majority of studies on CSR and RL systems show that firms incorporate CSR principles on forward logistics and supply chain management, while there are no studies on RL and CSR practices. Finally, the majority of those models are based on strategic and decision support mathematical systems, which help the choice and implementation of the optimum set of environmental management practices without however, evaluating the performance of the current

RL practices [12]. Additionally, there has been little research on CSR quality issues and RL procedures.

Therefore, this chapter aims to develop a holistic framework that includes a set of CSR quality indicators which can be used to evaluate the social responsibility quality of RL. In particular, this chapter suggests the introduction of CSR principles in the RL procedures and develops a framework of indicators based on the triple bottom line (TBL) approach. The framework focuses on the global reporting initiative (GRI) guide and the literature on RL. The contribution of this chapter is to investigate the prospects of a quality performance evaluation framework based on the relationship of RL to TBL.

Subsequently, a literature review of relevant environmental and sustainable performance frameworks associated with RL systems is presented. This is followed by an analysis of CSR principles (Sect. 4.2), the TBL concept, and GRI indices as well as by the presentation of current RL practices. Section 4.3 focuses on the development of a framework for evaluating the CSR performance of RL based on GRI and finally, in Sect. 4.4 the discussion and concluding remarks are outlined.

4.2 Logistics and Socioenvironmental Quality Topics: A Short Review

This section includes a review of various phases of the overall supply chain management and socioenvironmental quality topics. The wider scope of this review is the analysis and clarification of the current academic landscape in order to describe the precise position of RL in the overall supply chain management literature. It is also aimed to identify similar examples of implementation of social and environmental quality topics in other phases of the supply chain (e.g., forward supply chains), which may be useful for improving the socioenvironmental quality performance of RL. To this end, the relevant literature can be classified into the following general categories: (a) supply chain management (forward activities), (b) RL (backward activities), and (c) closed-loop supply chain (CLSC) management (forward and backward activities). The first category is examined to identify similar successful examples of general supply chain management, while the second category is related to the main objective of this chapter. Finally, the third category is studied because the majority of the examined models provide findings regarding the relationship between supply chain management and RL procedures; this fact facilitates the wider scope of this chapter.

Moreover, environmental and social management issues are also examined, as well as the ways in which such issues are associated with the three types of supply chain management mentioned previously. Furthermore, the presented literature is divided into two general categories: theoretical and empirical. The framework of the analysis that we developed is presented schematically in Fig. 4.1.

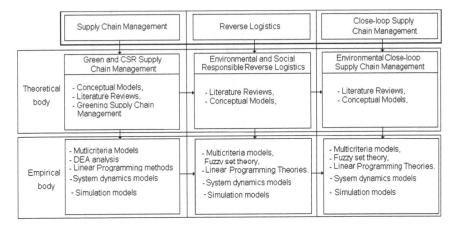

Fig. 4.1 A framework for the analysis of the literature

4.2.1 Theoretical Body of Literature

It refers to green supply chain management. The supply chain management encompasses all activities associated with the transformation and flow of goods and services, as well as information flows from the source of materials. Walker et al. [13] identify a range of drivers and barriers faced by the public and private sector in their effort to implement environmental management quality practices in order to make their supply chain management more environmentally friendly. Schmidt and Schwegler [14] suggest a set of ecological indicators for measuring the environmental and ecological performance of supply chain based on the eco-efficiency and life cycle concepts. Tsoulfas and Pappis [15] propose a model for the environmental performance of supply chains, while Wu and Dunn [16] suggest a conceptual model to explain the ways in which a firm can green supply chain management. Srivastrava [17] provides a literature review about green supply chain management.

Similarly, other researchers provide theoretical and conceptual frameworks for incorporating green and social issues into RL activities. RL include a complete process of planning, implementing, and controlling efficiently and effectively the flow of raw materials and inventory process of finished goods in the site of firms. In this context, Kumar and Putnam [18] develop an RL model in order to recover, recycle, reuse, and remanufacture end-of-life product materials. They identify that there are several drivers that encourage firms to implement environmental strategies to RL such as market competition, regulations, and globalizing growth. Similarly, they identify that supply chain coordination for the redesign of products, disassembly methods, and services improve the economic position and general quality profile of firms. Sarkis et al. [19] incorporate the social dimension along with the economic and the environmental, into RL practices. They note that social issues are associated with external population, human capital, productive capital, community capital, stakeholder participation, and socioeconomic performance.

The final part of this literature review includes research associated with CLSC management and environmental management practices. CLSC management embraces both forward and backward supply chain management actions. Kumar and Malegeant [20] propose a model to assist trade-offs between a manufacturer and an eco-non-profit organization in the collection process of a CLSC, stating that this process is a good strategy that benefits the manufacturer. Furthermore, Pagell et al. [21] present a framework in order to highlight the supply chain side of recycling, including various recycling options and other strategic implications.

4.2.2 Empirical Body of Literature

This body of literature includes academic work regarding green and CSR issues in supply chain, RL, and CLSC. According to the first category, a number of academics propose multi-criteria, linear programming methods, and simulation models to assist firms in making decisions according to environmental management practices in supply chain management. In this context, Sarkis [22] proposes a dynamic nonlinear multi-variate decision model for decision-making within the green supply chain. Lee et al. [23] provide a hybrid method based on the analytic hierarchy process (AHP) technique and a fuzzy set theory to help firms to select green suppliers. Additionally, Hymphreys et al. [24] develop a decision support tool which helps firms to incorporate environmental criteria into their supplier selection process. Similarly, various multi-criteria decision-making methods are proposed for modeling and analyzing supply chain networks with CSR. Finally, Presley et al. [25] propose a methodology for the decision-making of manufacturing in organizational product and process innovation stages by using soft system theory.[1]

Similarly, some academics propose linear and nonlinear methods to assist firms in selecting essential management practices associated with RL. Kannan et al. [26] provide a multi-criteria group decision-making model in fuzzy environment in order to help the decision-making process of firms regarding alternative environmental management practices in RL. Presley et al. [7] suggest an activity-based management methodological framework to frame decisions using corporate sustainability. In particular, this framework provides analytical steps in order to introduce TBL in RL procedures and mutli-criteria decision-making technique to

---

[1] Soft systems methodology (SSM) aims to investigate the cases when systems engineering methodology could solve management situations. SSM is considered more as a learning system, or, in other words, a process for acquiring knowledge about the parameters of a system.

help the selection of firms according to different alternative sustainability strategies. Furthermore, the proposed model also includes operational impact analysis focusing on tactical and strategic impacts, while their methodology also combines many existing practices such as activity-based costing and the balanced scorecard. Efendigil et al. [27] offer a model for assisting firms in determining the essential third-party RL based on neural networks and fuzzy logic. The proposed methodology considers that conventional factors such as price, quality, and flexibility are very limited for modern businesses. Thus, Efendigil et al. [27] incorporate various environmental factors in the design of RL since they could play an important role in enhancing profit maximization and quality issues. Neto et al. [28] propose a new CLSC management by incorporating environmental hot spots in the electric and electronic equipment supply chain of firms. Gupta and Evans [29] provide an optimization model that assists the decision-making of members of CLSC. Additionally, Kumar and Yamaoka [30] analyze supply chain design for the Japanese car industry by using system dynamics and modeling analysis of the closed loop, including relationships between reduction, reuse, recycling, and disposal. Georgiadis and Besiou [31] propose a system dynamics model to assess sustainability issues in order to assist manufacturing in managing effectively, in environmental and economical terms, their CLSC. Furthermore, Georgiadis and Vlachos [32] propose a dynamic model of a CLSC for product recovery. They consider that incorporating economic and environmental issues could be an important driving force for the development of a well-organized CLSC management. Thus, they examine the impact of environmental issues on the long-term behavior of a single product supply chain with product recovery.

4.3 Framework for Evaluating the CSR quality of RL procedures

This section describes the basic steps in the development of the proposed framework. In order to present the methodological approach, Fig. 4.2 has been developed. The first step includes the selection of the RL procedures based on the recent literature on RL. The second step includes the selection of the quality indicators based on the current literature on GRI, CSR, and TBL. The type of such quality indicators will be associated with the financial mode, the environmental mode, and the social one. The third step contains the development of the social responsibility quality of RL matrix and, the fourth step encompasses the presentation of the most important phases for using this methodology, namely sample selection, RL procedures selection, determining SR quality, RL matrix, and analysis of results.

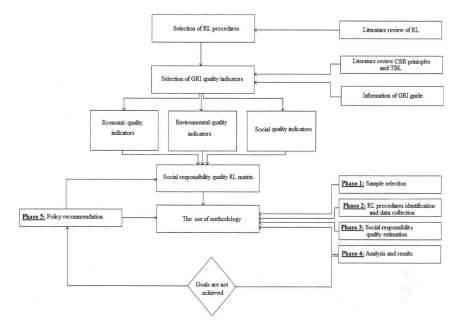

Fig. 4.2 The structure of the proposed framework

4.3.1 RL Procedures

Although many researchers focus on the study of RL, there is no consensus regarding its concept [34, 35]. Various definitions are proposed, such as reverse channels, reverse flow, recycling, reuse, and remanufacturing. Dowlatshahi [33] defines RL as

> a process by which a manufacturing entity systematically takes back previously shipped products or parts from the point-of-consumption for possible recycling, remanufacturing or disposal

(p. 1361). Similarly, Hu et al. [11] define RL as

> the process of logistics management involved in planning, managing, and controlling the flow of wastes for either reuse or final disposal of wastes

(p. 457). Serrato et al. [34] consider

> reverse logistics to include all activities associated with collecting, inspecting, reprocessing, redistributing, and disposing of items after they were originally sold

(p. 4289). However, the most well-known definition of RL is given by the European working group on RL as the process of planning, implementing, and controlling flows of raw materials, in process, inventory, and finished goods, from a manufacturing, distribution or use point, to a point of recovery or point of proper disposal.

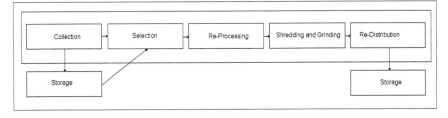

Fig. 4.3 RL Framework [39]

As presented in the previous section, a range of studies have developed quantitative models, case studies, and theory research questions. Moreover, a number of RL models are proposed that could be classified into two general categories: deterministic and stochastic. The first category includes mixed integer linear programming models for recycling and construction waste. The second category includes stochastic models that take into account uncertainty. Some of these models are based on the maximization of net profits and others on maximizing other parameters, e.g., a firm's profile. In this sense, case studies have been conducted in different industrial sectors such as the shoe industry, the automotive industry, the pharmaceutical industry, and the chemical industry [35]. Some researchers, such as Dowlatshahi [36], build theoretical foundations for RL procedures that provide a strategic framework for assisting firms to design and implement remanufacturing operations in RL. Some other models focus on examining the recovery of products such as metal scrap brokers and glass recycling. These models assess techniques for reusing, recycling, and remanufacturing recovery products. Fleishmann et al. [37] mention that manufacturers adopt recovery products strategies as a result of three factors, i.e., environmental legislation (e.g., 2002/96/EC and 2003/108/EC), economic incentives (e.g., improved productivity), and consumer demand (e.g., increased market share).

Given this literature framework, a number of proposed models for describing either the overall procedures or specific procedures of RL are presented. Pokharel and Mutha [38] present a number of stages of a typical RL system, including input procedures (used products—new parts or modules), structure, and processes (collection, processing, remanufacturing, inspection, and consolidation), outputs (recycled material, spare parts, remanufactured products, and waste disposal). Fleischmann et al. [39] provide an integrated framework for production planning with the reuse of old products, including four steps: the first step is either the disassembly of old products or material recycling or the selling of old products for external reuse, the second step is material recycling, the third step is reassembly, and the final step is the production of new products. Summarizing, some standard procedures of an RL framework are presented in Fig. 4.3.

4.3.2 CSR Principles and Triple Bottom Line

Some academics have suggested that firms should introduce a range of ethical and quality issues in their day-to-day business. In particular, the debate has focused on many definitions and practices in order to assist in introducing the general principles of CSR, corporate sustainability, and TBL. Nowadays, many authors claim that these innovative concepts can improve the production and operational procedures of firms. More specifically, Castka and Balzarova [40] find that the quality field can be associated with the CSR agenda. However, traditional operational and production processes should be reinvented and renewed in key areas such as management systems, strategic management, operations, technology, CSR, and quality, and improvements in third-party certification and internal auditing practices. Similarly, Kok et al. [41] propose that the concept of CSR audit should be introduced in the current quality award/excellence models.

The aforementioned proposed combination of concepts is usually found in the literature on corporate environmental management. For example, Carriga and Mele [42] and van Merrewijk [43] present a range of definitions and management practices as well as discuss the similarities and differences between CSR, corporate sustainability, and TBL definitions. This variety of definitions is highlighted by the diversity of concepts such as corporate citizenship, business ethics, corporate social responsiveness, corporate sustainability, eco-efficiency, social performance, and CSR. Today, those definitions usually include ethical, social, environmental, and economic issues such as marketing, management, public responsibility, stakeholder management, social performance, and environmental performance.

Although there are various definitions of sustainability concepts, most academics agree that they have three basic and separate dimensions in common, namely the economic, environmental, and social ones. The first dimension is related to the financial contribution of firms to stakeholders such as shareholders, employees, and community (e.g. GDP). The second dimension refers to internal environmental management aspects and external natural resources conservation issues. Finally, the third dimension includes quality and ethical issues regarding risk management issues, health and safety issues, and employment issues.

4.3.3 The GRI Approach

The GRI is a framework for assisting firms with sustainability issues performance through specific report standards. The GRI guidelines provide standards particularly for recording on an annual basis the economic, environmental, and social performance, operations, products, and services of firms. The purpose of such guidelines is mainly to develop globally acceptable reporting guidelines that are appropriate for each type of industry (as well as the different cultural context of

Table 4.1   GRI performance indicators [50]

Performance indicators	Core indicators	Additional indicators	Total
Economic	10	3	13
Environmental	16	19	35
Social	24	25	49
Total	50	47	97

countries) such as manufacturing companies (e.g. electronics, automobiles, textiles, plastic, and minerals) and services (e.g. banks and tourist companies). These guidelines are of a voluntary character and provide flexibility to firms in deciding the extent to which financial and nonfinancial information should be disclosed. Furthermore, these guidelines provide the basic steps for identifying the contribution of a firm to sustainable development or, in other words, measuring the level of CSR.

This approach stems from the new trend—which is supported by various authors—to associate the concept of total quality management (TQM) with the concepts of sustainability and CSR. Caska et al. [44] highlight that issues associated with CSR assist firms in achieving quality, sustainability, and stakeholders' goals. Caska and Balzrova [40] propose a framework for analyzing synergies of quality management and CSR issues, while MacAdam and Leonard [45] identify some important key issues in order for CSR and sustainability issues to be incorporated effectively into the daily operations of firms through TQM procedures.

The GRI is based on the TBL approach and includes a separate section with performance indicators including economic, environmental, and social indicators. The TBL approach was initially established by Elkington [46] who considered that the business community should meet three goals of modern economy: economic growth, environmental preservation, and social equity. Thus, a number of methodologies have been developed for measuring the TBL of firms. Such an indicative methodology is the GRI, which has provided a number of relevant indicators. Table 4.1 shows the performance indicators of GRI. In general, performance indicators are classified into two categories: core and additional ones. Oskarsson and von Malmborg [47] propose that quality issues are part of the concept of sustainable development, while Isaksson [48] considers GRI guidelines to be a good vehicle, but with some limitations, for measuring quality issues. However, there are several reasons for using these guidelines such as transparency, inclusiveness, auditability, accountability and the fact that they are generally accepted.

The economic indicators proposed by GRI are based on the value added statement scheme. Specifically, they measure the effect of a firm's operation on the financial positions of stakeholders. Isaksson [48] highlights some economic indicators of GRI guidelines which are associated with quality issues such as monetary flow indicators, product value, value added, and cost of poor product quality.

However, Kok et al. [41] consider that business ethical and sustainability performance are very important for measuring business excellence and quality issues. Karapetrovic and Willborn [49] claim that quality and responsibility issues of business go beyond traditional issues associated with the quality of products and services. They go toward modern issues of environmental management, health and safety, production and operations management issues. To this end, GRI provides a range of environmental indicators which cover areas such as waste management reduction, elimination of influences on biodiversity, and emissions impact minimization. Overall, this category includes 35 indicators (16 core indicators and 19 additional ones). Similarly, the social dimension includes 49 indicators such as quality of management, health and safety, wages and benefits, equal opportunities policy, training/education, child labor, forced labor, freedom and association, human rights (including indigenous and security), suppliers and products, and services [50]. The presentation of economic, environmental, and social indicators of GRI is made in the following section.

4.3.4 The Matrix of Social Responsibility Quality of RL

This section presents a conceptual framework aiming at developing a set of indicators to help firms and other stakeholders to measure the social responsibility performance of an RL system. In Fig. 4.4 we present the conceptual model that we designed in which social responsibility quality indicators are developed. In particular, in Fig. 4.4, we indicate the relationship between the GRI standards and the RL procedures. This association facilitates firms to understand and evaluate their social responsibility quality status. The first part of Fig. 4.4 represents the main sections of the GRI guidelines which include:

• direct economic impacts on customers, suppliers, employees, providers of capital,
• environmental impacts on biodiversity, natural resources, energy and
• social impacts on labor, human rights, society, and product responsibility.

These sections of GRI are associated with the common steps of RL, namely collection, selection, reprocessing, shredding and grinding, and redistribution (as shown in Fig. 4.2 and in the second part of Fig. 4.4). The final part of Fig. 4.4 shows the relationship between the social responsibility quality concept and RL which is the base for preparing the necessary indicators to measure the overall social responsibility performance of RL systems. It should be pointed out that from the total number of indicators mentioned in Table 4.1 we utilize only 57 for developing the social responsibility quality of RL matrix, which is associated with economic, environmental, and social quality. Specifically, we use 9 economic, 16 environmental, and 32 social indicators.

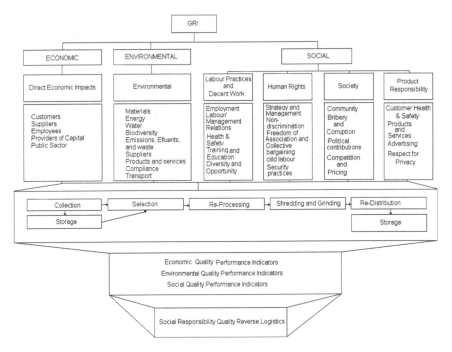

Fig. 4.4 Conceptual model of social responsibility quality of RL

4.3.4.1 Indicators of Economic Quality

The economic aspect is one of the three basic pillars for measuring the social responsibility quality of RL. For its measurement various indicators have been proposed. For example, Weeks et al. [51] propose various financial indicators for RL such as machine costs, labor costs, material costs, capital stock, costs to retool machine, and depreciation. However, GRI guidelines provide some indicators which are appropriate for measuring mainly quality issues in the overall operation of a firm. Some of these indicators are also essential for measuring the economic efficiency of RL. Some direct economic indicators are the following:

(a) net sales of reuse, resalable, and recyclable products (EC_1),
(b) costs of used and returned materials (EC_2),
(c) total payroll and benefits for staff in RL procedures (EC_3),
(d) distribution to providers of capital broken down by interest on debt and borrowings for RL procedures (EC_4),
(e) taxes paid broken down by country associated with RL procedures (EC_5),
(f) subsidies associated with RL (EC_6),
(g) donations to community, civil society, and other groups associated with RL (EC_7),
(h) total spent on noncore business infrastructure development associated with RL systems (EC_8), and

(i) indirect economic impacts regarding RL systems (EC_9).

4.3.4.2 Indicators of Environmental Quality

Today, some academics identify the relationship and potential impact of an RL system on the natural environment [15]. For example, they focus on examining the drivers that press a firm to recycle and remanufacture products. To this end, a range of indicators have been suggested, such as the design for disassemble index (DfDI), the integrated disassemble and recycling score (IDRS), as well as life cycle analysis models and environmental management performance systems.

Under GRI guidelines, the most appropriate environmental core indicators for RL are the following:

(a) total materials uses (EN_1),
(b) percentage of waste materials (EN_2),
(c) direct energy uses in the procedures of RL (EN_3),
(d) indirect energy use in the procedures of RL (EN_4),
(e) total water use in the procedures of RL (EN_5),
(f) total amount of waste in the procedures of RL (EN_6),
(g) significant discharges to water in the procedures of RL (EN_7),
(h) significant environmental impacts of products and services in the procedures of RL (EN_8),
(i) percentage of the weight of products sold at the end of the products' useful life (EN_9),
(j) incidents of and fines for noncompliance in the procedures of RL (EN_{10}).
(k) Other indirect energy use (EN_{11}),
(l) business units currently operating around protected or sensitive areas (EN_{12}),
(m) other relevant indirect greenhouse gas emissions (EN_{13}),
(n) all production and transport of any hazardous waste (EN_{14}),
(o) environmental performance of suppliers (EN_{15}), and
(p) the significant environmental impacts of transportation used for logistical purposes (EN_{16}).

4.3.4.3 Indicators of Social Quality

A limited amount of academic work aims to evaluate the social performance of RL procedures and provide examples from practice of the effects of RL activities. To this end, Sarkis et al. [19] propose some social indicators in RL such as employment stability, employment practices, health and safety, human capital, productive capital, community capital, and stakeholder influence. Similarly, Halldorsson et al. [52] provide the following social indicators for RL systems: reduced traffic congestion, education in energy saving, driving automation of loading and unloading, respecting driving and resting time rules.

Based on the GRI guidelines, there are four categories of indicators regarding social responsibility issues. The first category includes core and additional labor indicators regarding RL including:

(a) breakdown of workforce in RL procedures(LA_1),
(b) net employment creation in RL procedures (LA_2),
(c) compliance with the International Labor Organization (ILO) Code in RL procedures (LA_3),
(d) health & safety committees in RL procedures (LA_4),
(e) average hours of training in RL procedures (LA_5),
(f) employee benefits beyond those legally mandated in RL (LA_6),
(g) provision for formal worker representation in management of RL (LA_7),
(h) compliance with the ILO guidelines for occupational health management systems in RL (LA_8),
(i) programs to support the continued employability of employees (LA_9),
(j) programs for skills management or for lifelong learning (LA_{10}).

The second category includes human resources indicators based on GRI adjusted to RL systems. More specifically:

(a) policies to deal with all aspects of human rights in RL (HR_1),
(b) policies to evaluate human rights performance within the supply chain (HR_2),
(c) global policy preventing all forms of discrimination in operations in RL (HR_3),
(d) policy prohibiting child labor in RL procedures (HR_4),
(e) policy to prevent forced and compulsory labor in RL procedures (HR_5),
(f) employee training on practices concerning human rights in RL procedures (HR_6),
(g) practical human rights issues in RL procedures (HR_7),
(h) confidential nonretaliation policy for the employee (HR_8),
(i) policies to address the needs of indigenous people in RL procedures (HR_9),

The third category refers to issues related to society. Such indicators are as follows:

(a) policies to manage impacts on communities in areas affected by RL activities (SO_1),
(b) awards received relevant to social, ethical, and environmental performance of RL (SO_2),
(c) the policy on customer health and safety (SO_3).

The fourth category is about product responsibility issues. Some representative indicators of this category are:

(a) policies on customer health and safety during use of products (PR_1),
(b) management systems related to product information and labeling (PR_2),
(c) procedures for consumer privacy (PR_3),
(d) types of instances of noncompliance with regulations concerning customer health and safety (PR_4),

(e) the number of complaints upheld by regulatory or similar official bodies (PR_5),
(f) voluntary code compliance, product labels, or awards with respect to social and environmental responsibility (PR_6),
(g) the types of instances of noncompliance with regulations concerning product information (PR_7), policies related to customer satisfaction (PR_8),
(h) procedures for adherence to standards related to advertising (PR_9),
(i) types of breaches of advertising and marketing regulations (PR_{10}).

4.3.4.4 Social Responsibility Quality of RL Matrix

Table 4.2 indicates how previous categories of GRI indicators are classified into different RL procedures in order to assist the evaluation of CSR quality performance of such procedures. On the one hand, some standard procedures of an RL system, as presented in Sect. 4.3.1, are recorded in the columns of Table 4.2. On the other hand, the rows of Table 4.2 demonstrate the types of indicators of the TBL approach, i.e., economic, environmental, and social indicators. Each cell of Table 4.2 corresponds to the types of indicators which could be measured at least in some stages/procedures of an overall RL system. For example, economic indicators could be classified as follows:

- EC_6, EC_7, EC_9 could be measured at the collection stage,
- EC_9 at the selection stage,
- EC_2, EC_3 at the reprocessing stage,
- EC_4, EC_8, EC_9 at the shredding and grinding stage and, finally,
- EC_1, EC_5 could be measured at the redistribution stage.

Similarly, environmental and social indicators could be classified accordingly.

4.3.5 The Use of the CSR Quality Framework

In what follows we provide the theoretical steps of the methodology presented previously through the social responsibility RL matrix and the relevant indicators. More specifically, this section describes all the phases of the proposed framework for implementing this methodology in order for the CSR quality of RL to be estimated. The proposed framework includes five phases:

(a) Phase 1: sample selection

This is an initial phase that includes the identification of the sample where the social responsibility quality of RL matrix should be determined. The basic characteristics of the potential firms should be the implementation of CSR, sustainability, and environmental strategies along with strategies on RL. In order for the potential firms of the sample to be selected, other criteria should be the type of sector in which they operate and their size, i.e., small, medium-sized, or large enterprises.

Table 4.2 Social responsibility RL matrix[a]

Social responsibility aspects	Social responsibility indicators	Collection	Selection	Reprocessing	Shredding and grinding	Redistribution
(1)	(2)	(3)	(4)	(5)	(6)	(7)
Economic	EC_i, $i = 1,\ldots,9$	EC_6, EC_7, EC_9	EC_9	EC_2, EC_3	EC_4, EC_8, EC_9	EC_1, EC_5
Environmental	EN_j, $j = 1,\ldots,16$	EN_2, EN_3, EN_6, EN_9, EN_{11}, EN_{12}, EN_{15}	EN_3, EN_6, EN_{10}, EN_{11}, EN_{12}	EN_1, EN_4, EN_5, EN_7	EN_3, EN_7	EN_3, EN_8, EN_{13}, EN_{14}, EN_{15}, EN_{16}
Social						
Labor indicators	LA_k, $k = 1,\ldots,10$	LA_1, LA_3, LA_5, LA_6, LA_7, LA_8, LA_9, LA_{10}	LA_1, LA_5, LA_6, LA_8, LA_9, LA_{10}	LA_1, LA_5, LA_6, LA_8, LA_9, LA_{10}	LA_1, LA_3, LA_4, LA_5, LA_6, LA_8, LA_9	LA_1, LA_2, LA_3, LA_5, LA_6, LA_8, LA_9, LA_{10}
Human resources	HR_l, $l = 1,\ldots,9$	HR_1, HR_2, HR_4	HR_2, HR_3, HR_4, HR_6, HR_7, HR_8, HR_9	HR_1, HR_2, HR_3, HR_5, HR_6, HR_7, HR_8	HR_1, HR_2, HR_6, HR_7, HR_8	HR_1, HR_2, HR_6, HR_7, HR_8, HR_9
Society	SO_m, $m = 1,\ldots,3$			SO_1.		SO_2, SO_3.
Product responsibility	PR_n, $n = 1,\ldots,10$			PR_3, PR_6, PR_7		PR_1, PR_2, PR_3, PR_4, PR_5, PR_6, PR_8, PR_9, PR_{10}, PR_{11}

[a] The most appropriate indicators of the total number of proposed indicators of GRI guidelines for measuring the CSR quality of RL procedures

(b) Phase 2: RL procedures identification and data collection

In the second phase, an initial analysis should be made assessing the comprehensiveness of RL procedures. The selected firms should be classified into different categories according to the number and type of their procedures. For instance, there should be a separation between those firms that have implemented a complete plan for RL and those that adopt only a number of procedures. Additionally, the data should be drawn from current databases such as corporate social CSR reports, sustainability reports, and environmental reports as well as information from websites in relation to CSR issues. Relevant information will be also identified by interviews and questionnaire-based research

(c) Phase 3: social responsibility quality estimation

In this phase, the collected data of the previous phase should be used to fill in the social responsibility RL matrix for each firm separately. The social responsibility RL matrix provides the economic, environmental, and social quality picture of each of the procedures of the RL of selected firms.

(d) Phase 4: analysis of results

In this phase, the result of the estimated indicators of social responsibility matrix should be analyzed. First, the weaknesses and challenges of the current strategies implemented for any RL procedures of each firm examined should be identified. In case the goals of social responsibility quality are achieved, then the results should be published. On the contrary, if the goals of a firm have deteriorated as far as previous years or other firms of the sector are concerned, then some policy recommendation will be designed (Phase 5).

(e) Phase 5: policy recommendation

This phase includes the procedures for correcting the weaknesses and problems that have arisen in phase 4, as well as the policy recommendations for the ongoing improvement of CSR of RL. In particular, policy recommendation should encompass issues about the overall research from the sample selection phase to policy recommendation phase.

4.4 Conclusions and Future Research

This chapter proposes a new methodological framework with the aim to assist in evaluating the social responsibility quality of RL. It is based on the TBL approach and specifically on the GRI guidelines, and provides some series of mathematical indicators for evaluating the overall economic, environmental, and social responsibility quality of an RL system.

Moreover, the developed framework intends to make clearer and more comprehensive the limited number of present studies about CSR and RL issues. At the same time, it aims at overcoming the weaknesses of the majority of the present models, which focus either on financial or environmental aspects only, by evaluating the overall social responsibility performance. It also aims at contributing to the current literature by providing some standard and broadly accepted indicators based on the GRI guidelines.

It is important to note that the proposed standard indicators aim at facilitating firms or other interested organizations to attend the longitudinal improvement in the CSR performance of a specific firm in RL procedures or to make comparisons among different firms of different industries.

The chapter also attempts to highlight areas for further research beyond theoretical models related to TQM issues, in order to select specific CSR characteristics of these industries. The proposed framework could provide a strong theoretical foundation to balance economical, environmental, and social issues as well as the basis for a range of high-quality empirical studies while also being the springboard for future research in RL and CSR.

References

1. Ravi V, Shankar R (2005) Analysis of interactions among the barriers of reverse logistics. Technol Forecast Soc Chang 72:1011–1029
2. Gonzalez-Torre PL, Adenso-Diaz B, Artiba H (2004) Environmental and reverse logistics policies in European bottling and packaging firms. Int J Prod Econ 88:95–104
3. Subramanian R, Talbot B, Gupta S (2010) An approach to integrating environmental considerations within managerial decision-making. J Ind Ecol 14(3):378–398
4. Gottberg A, Morris J, Pollard S, Mark-Herbert C, Cook M (2006) Producer responsibility, waste minimization and the WEEE directive: case studies in eco-design from the European lighting sector. Sci Total Environ 359:38–56
5. Tsoulfas GT, Pappis CP, Minner S (2002) An environmental analysis of the reverse supply chain of SLI batteries. Resour Conserv Recycl 36:135–154
6. Georgiadis P, Vlachos D (2003) Analysis of the dynamic impact of environmental policies on reverse logistics. Oper Res Int J 3(2):123–135
7. Presley A, Meade L, Sarkis J (2007) A strategic sustainability justification methodology for organizational decisions: a reverse logistics illustration. Int J Prod Res 45(18–19):4595–4620
8. Ciliberti F, Pontrandolfo P, Scozzi B (2008) Logistics social responsibility: standard adoption and practices in Italian companies. Int J Prod Econ 113:88–106
9. Malony MJ, Brown ME (2006) Corporate social responsibility in the supply chain: an application in the food industry. J Bus Ethics 68:35–52
10. Teunter RH (2001) A reverse logistics valuation method for inventory control. Int J Prod Res 39(9):2023–2035
11. Hu TL, Sheu J-B, Huang K-H (2002) A reverse logistics cost minimization model for the treatment of hazardous wastes. Transp Res Part E 38:457–473
12. Ravi V, Shankar R, Tiwari MK (2005) Analyzing alternatives in reverse logistics for end-of-life computers: ANP and balanced scorecard approach. Comput Ind Eng 48:327–356
13. Walker J, Di Sisto L, McBain D (2008) Drivers and barriers to environmental supply chain management practices: lessons from the public and private sectors. J Purch Supply Manag 14:69–85
14. Schmidt M, Schwegler R (2008) A recursive ecological indicator system for the supply chain of a company. J Clean Prod 16:1658–1664

15. Tsoulfas GT, Pappis CP (2008) A model for supply chains environmental performance analysis and decision making. J Clean Prod 16:1647–1657
16. Wu H-J, Dunn SC (1995) Environmental responsible logistics systems. Int J Phys Distrib Logist 25(2):20–38
17. Srivastrava SK (2007) Green supply-chain management: a state-of-the-art literature review. Int J Manag Rev 9(1):53–80
18. Kumar S, Putnam V (2008) Cradle to cradle: reverse logistics strategies and opportunities across three industry sectors. Int J Prod Econ 115:305–315
19. Sarkis J, Helms MM, Hervani AA (2010) Reverse logistics and social sustainability. Corp Soc Responsib Environ Manag 17 (6):337–354
20. Kumar S, Malegeant P (2006) Strategy alliance in a closed-loop supply chain, a case of manufacturer and eco-non-profit organization. Technovation 26:1127–1135
21. Pagell M, Wu Z, Murthy NN (2007) The supply chain implications of recycling. Bus Horiz 50:133–143
22. Sarkis J (2003) A strategic decision framework for green supply chain management. J Clean Prod 11:397–409
23. Lee AHI, Kang H-Y, Hsu C-F, Hung HC (2009) A green supplier selection model for high-tech industry. Expert Syst Appl 36:7917–7927
24. Humphreys PK, Wong YK, Chan FTS (2003) Integrating environmental criteria into the supplier selection process. J Mater Process Technol 138:349–356
25. Presley AR, Sarkis J, Liles DH (2000) A soft systems methodology approach for product and process innovation. IEEE Trans Eng Manag 47(3):379–392
26. Kannan G, Pokharel S, Kumar PS (2009) A hybrid approach using ISM and fuzzy TOPSIS for the selection of reverse logistics provider. Resour Conserv Recycl 54:28–36
27. Efendigil T, Onuut S, Knogar E (2008) A holistic approach for selecting a third-party reverse logistic provider in the presence of vagueness. Comput Ind Eng 54:269–287
28. Neto JQF, Wlater G, Bloemhof J, van Nunen JAEE, Spengel T (2009) From closed-loop to sustainable supply chains: the WEE case. Int J Prod Res 1–19
29. Gupta A, Evans GW (2009) A goal programming model for the operation for the operation of closed-loop supply chains. Eng Optim 41(8):713–735
30. Kumar S, Yamaoka T (2007) System dynamics study of the Japanese automotive industry closed-loop supply chain. J Manuf Technol Manag 18(2):115–138
31. Georgiadis P, Besiou M (2010) Environmental and economical sustainability of WEEE closed-loop supply chains with recycling: a system dynamics analysis. Int J Adv Manuf Technol 47:475–493
32. Georgiadis P, Vlachos D (2004) The effect of environmental parameters on product recovery. Eur J Oper Res 157:449–464
33. Dowlatshahi S (2010) A cost-benefit analysis for the design and implementation of reverse logistics systems: case studies approach. Int J Prod Res 48(5):1361–1380
34. Serrato MA, Ryan SM, Gaytan J (2007) A Markov decision model to evaluate outsourcing in reverse logistics. Int J Prod Res 45(18):4289–4315
35. Fleischmann M, Krikke HR, Dekker R, Flapper SDP (2000) A characterization of logistics networks for product recovery. Omega 28:653–666
36. Dowlatshahi S (2005) A strategic framework for the design and implementation of remanufacturing operations in reverse logistics. Int J Prod Res 43(16):3455–3480
37. Fleischmann M, Beullens P, Bloemhof-Ruwaard J, van Wassenhove L (2001) The impact of product recovery on logistics network design. Prod Oper Manag 10(2):156–173
38. Pokharel S, Mutha A (2009) Perspectives in reverse logistics: a review. Resour Conserv Recycl 53:175–182
39. Fleischmann M, Bloemhof-Ruwaard JM, Dekker R, van der Laan E, van Nunen JAEE, van Wassenhove LN (1997) Quantitative models for reverse logistics: a review. Eur J Oper Res 103:1–17

40. Castka P, Balzarova MA (2007) A critical look on quality through CSR lenses: key challenges stemming from the development of ISO 26000. Int J Qual Reliab Manag 24(7):738–752
41. Kok P, van der Wiele T, McKenna R, Brown A (2001) A corporate social responsibility audit with a quality management framework. J Bus Ethics 31:285297
42. Carriga E, Mele D (2004) Corporate social responsibility theories: mapping the territory. J Bus Ethics 53:51–71
43. van Marrewijk M (2003) Concepts and definitions of CSR and corporate sustainability: between agency and communion. J Bus Ethics 44:95–105
44. Castka P, Bamber CJ, Bamber DJ, Sharp JM (2004) Integrating corporate social responsibility of a feasible CSR management system framework. TQM Mag 16(3):216–224
45. McAdam R, Leonard D (2003) Corporate social responsibility in a total quality management context: opportunities for sustainable growth. Corp Gov 3(4):36–45
46. Elkington J (1998) Cannibals with the forks: the triple bottom line of 21st century business. New society Publishers, Gabriola Island
47. Oskarsson K, von Malmborg F (2005) Integrated management systems as a corporate response to sustainable development. Corp Soc Responsib Environ Manag 12:121–128
48. Isaksson R (2006) Total quality management for sustainable development process based system models. Bus Process Manag J 12(5):632–645
49. Karapetrovic S, Willborn W (1998) Integration of quality and environmental management systems. TQM Mag 10(3):204–213
50. Dixon RD, Mousa GA, Woodhead A (2005) The role of environmental initiatives in encouraging companies to encourage in environmental reporting. Eur Manag J 23(6):702–716
51. Weeks K, Gao H, Alidaeec B, Rana DS (2010) An empirical study of impacts of production mix, product route efficiencies on operations performance and profitability: a reverse logistics approach. Int J Prod Res 48(4):1087–1104
52. Halldorsson A, Kotzab H, Skhott-Larsen T (2009) Supply chain management on the crossroad to sustainability: a blessing or a curse? Logist Res 1:83–94

Chapter 5
Quality Assurance and Consumer Electronics Recycling

Robert Sroufe

Abstract The information within this chapter examines the characteristics of consumer electronic recycling systems to show how quality assurance has evolved to meet the current needs of reverse logistics demanufacturers. A review of the literature reveals several trends regarding the amount of e-waste, recycling programs, the influence of international regulations, a focus on large-scale operations, and emerging recycling certifications. Given the dynamic context of consumer electronics recycling systems and opportunities for new competitive capabilities, information within this chapter provides exploratory field study insight from a small US recycling firm. A primary contribution of this chapter is found in filling a gap in the literature to advance our understanding how small firms are overcoming emerging challenges and taking advantage of opportunities facing them within reverse supply chains focusing on recycling of consumer electronics and information technology assets. The field study sheds new light on quality assurance through emerging standards, contemporary opportunities for emerging business models within the industry, and implications for the future of reverse logistics practices and research.

5.1 Introduction

The evolving role of reverse logistics and the expansion of quality assurance, to include transparent destruction and recycling of consumer electronics (CE) products and information technology (IT) assets, present a timely opportunity to

R. Sroufe (✉)
Duquesne University, 600 Forbes Avenue, Pittsburgh, PA 15238, USA
e-mail: sroufer@duq.edu

Y. Nikolaidis (ed.), *Quality Management in Reverse Logistics*,
DOI: 10.1007/978-1-4471-4537-0_5, © Springer-Verlag London 2013

examine emerging approaches to quality assurance. CE products include, but are not limited to TVs and other video equipment, computers, assorted peripherals, audio equipment, and phones, while IT assets include computers, routers, and network equipment. For the last 25 years the products of technological advancement and innovation have entertained the world, improved data processing, and placed people in immediate communication with each other. These products have clearly improved the quality of life for those in developed countries, but they were designed and built with materials that have an adverse impact on the environment. Additionally, these devices contain sensitive personal and corporate data needing careful management at the devices end of life [19]. While future devices may contain less toxic materials through a more sustainable design, the current devices contain toxins (e.g., lead, cadmium, chromium, mercury, and brominated flame retardants) that require more effort to dispose of properly. Constant and rapid innovation causes premature obsolescence for these products that are retired from service and vast amounts of sensitive data still left within products such as cell phones and computers. Some products can be refurbished and resold, but most of this material is destined to be discarded creating the fastest growing waste stream in the world. This growing waste stream should serve as a wake-up call to supply chain professionals and entrepreneurs wanting to better understand where to find new opportunities regarding the strategic importance of end of life management [44], supply chain management, reverse logistics, and quality assurance.

Generally speaking, supply chain management is the sequencing of an organization's facilities, functions, and activities involved in producing and delivering products and services from cradle to end customer. The sequencing begins with the extraction and supply of raw materials from the earth "cradle" and extends to customers, stopping short of landfills, and disposal at the product "grave". This long standing, one-way flow of goods from manufacturers to consumers, has overlooked the opportunities found within cradle to cradle or closed loop systems [13], end-of-life management, and life cycle information management systems [37]. Within these systems, both reverse logistics and quality management have dynamic and emerging relationships in the capture and recovery of value within CE products and IT assets. Despite this opportunity for value creation, there remains a dearth of information regarding the tactics and strategies available within reverse logistics or recycling systems, and the role of certification to existing quality standards.

Whereas supply chain management looks at the forward flow of products and services, reverse logistics involves "the process of planning, implementing, and controlling the efficient, cost effective flow of raw materials, in-process inventory, finished goods and related information from the point of consumption to the point of origin for the purpose of recapturing value or proper disposal" [7]. Reverse logistics management also includes topics such as disassembly, demanufacturing, general and hazardous waste treatment, recycling, and disposal, in addition to typical logistics activities such as information flow, transportation, and warehousing [18]. Basically, reverse logistics is the process of moving products from their typical final destination for the purpose of capturing value, or proper disposal.

For the purpose of this chapter, the reverse logistics process includes the management and the sale of materials from CE and IT assets after the useful life of the product. This "end user-to-cradle" approach to materials acquisition finds resources going at least one step back in the supply chain. For instance, goods move from the customer to a distributor, manufacturer, or recycler.

The motives for effective recycling systems to manage large, dynamic electronic waste (e-waste) flows are supported in the literature by supply chain management [33, 54], reverse logistics [26], and evolving certification programs [38, 53]. Interestingly, these motives combine to create a current paradox within the literature. The paradox involves the need for large-scale recycling systems with a focus on large firms to handle the amount of e-waste generated by developed countries while overlooking the development of recycling infrastructure through small demanufacturers. There is a gap in the literature regarding the role of quality assurance in e-waste recycling for which there is little empirical evidence and few, if any case studies. Both the paradox and research oversight provide a gap for which this chapter and an exploratory field study methodology are well positioned to fill.

Given that, the United States generate 10 million tons of waste electrical and electronic equipment a year, the United Kingdom generates 7 million tons, and Japan produces 5 million tons of this waste [54], studies within the US context provide insight as to current issues and opportunities within the country producing the most amount of e-waste. In the US, government, usually municipalities, plays the primary role in offering e-waste collection and recycling programs. Few states have mandated extended producer responsibility (EPR) recycling programs. A significantly higher percentage of programs are being operated by nongovernment entities, such as recyclers, and solid waste haulers [35]. This trend provides an opportunity for insight and anecdotal evidence of contemporary issues derived from exploratory field studies. To date, much of the extant literature has focused on large manufacturers issues with asset management, extended producer responsibility, and supply chain optimization. The challenges and opportunities faced by small demanufacturers present an opportunity to reveal hidden challenges and provide new opportunities for both practitioners and researchers in understanding the growing industry of e-waste recycling.

The literature review and information within this chapter helps to examine the characteristics of CE recycling system trends to uncover the emerging opportunities for business models within the industry. Through the lens of already established standards and certifications, readers will be able to see how quality assurance has evolved to meet the current needs of reverse logistics demanufacturers. This chapter progresses as follows, first is an examination of emerging trends regarding the amount of e-waste, recycling programs, the influence of international regulations, recycling certifications, and collection methods. Then comes a review of quality assurance through the lens of recognized international standards, before discussing field study methods. A gap in current research provides motivation for and an opportunity to draw insight form an exploratory field study. The second half of this chapter describes a field study of a small US recycling firm in Pittsburgh, Pennsylvania before looking at the operational, tactical, and strategic implications of the use of quality assurance standards and certifications. The information within this chapter sheds new light on the direction of quality assurance through emerging standards, implications for future business development, and opportunities for new research.

5.2 Emerging Trends

CEs—make up almost 2 % of the municipal solid waste stream in the US [49]. Although electronics comprise a small percentage of the total municipal solid waste stream, the quantity of e-waste that we are generating is steadily increasing. In 1998, the National Safety Council Study estimated about 20 million computers became obsolete in 1 year. Fast forward to 2007, that number has more than doubled according to USEPA's most recent estimates. Of the 2.25 million tons of US-based TVs, cell phones, and computer products ready for end-of-life (EOL) management in 2007, 18 % (414,000 tons) was collected for recycling and 82 % (1.84 million tons) was disposed of, primarily in landfills. From 1999 to 2005, recycling rate was relatively constant at about 15 %. During these years, the amount of electronics recycled increased but the percentage did not as the amount of electronics sent for EOL management increased each year as well. For 2006–2007, the recycling rate increased to 18 %, with the help of several states starting mandatory collection and recycling programs for electronics [50].

The useful life of CE devices is short, and decreasing as a result of rapid changes in equipment features and capabilities. This creates a growing waste stream of obsolete electronic equipment and e-waste [23]. Conventional disposal methods for CE and IT assets have economic and environmental disadvantages. As a result, waste management and recycling trends have provided new proving grounds for the impacts of quality assurance within reverse logistics. But, CE and IT asset recycling has a relatively short history with an emerging infrastructure, and a need for a greater understanding of the value proposition surrounding this form of logistics and the role of quality assurance. Key attributes of the CE and IT asset recycling space can be found in Fig. 5.1.

As with the recycling of other products, the establishment of appropriate infrastructure is essential to the successful implementation of CE recycling. Infrastructure determines the process methods and amounts of waste that can be processed. Infrastructure includes transportation, collection, recovery, and resale establishments. Recycling processes include collection, sorting, disassembly or demanufacturing, size reduction, and separation before the sale of commodities level materials. Opportunities for sale or reuse of parts can come in almost any stage, but are more common within disassembly and reduction. Factors that affect the recycling infrastructure are the amount of waste in the waste stream, the recycling technologies available, government regulations, and the economics of CE and IT asset products.

Fig. 5.1 Simplified flow diagram for recycling of CE and IT assets (modified from [23])

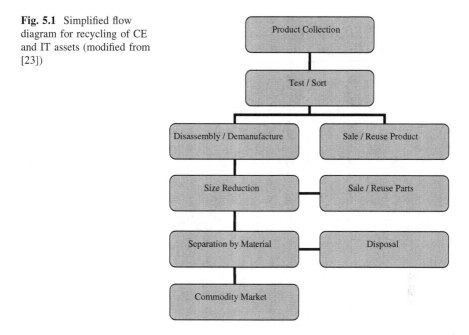

With CE recycling in its infancy, consumer and business recognition of the need for recycling is a critical factor to the further expansion of this industry. Many consumers do not recycle their electronic products when they first become defunct or obsolete. Some studies have shown that more than 70 % of retired CEs are kept in storage, typically for as many as 3–5 years [47]. Consumers falsely believe that these CEs have some value or do not know what to do with EOL items. With the development of electronic technologies, the residual value of outdated electronic devices decreases rapidly. Both the recovery value of parts and the machine resale value drop rapidly with the age of the machines. A computer's value approaches zero for machines with technologies greater than two generations old. Furthermore, old equipment is often more difficult to recycle than newer equipment. In general, old electronics waste contains a larger variety of materials, such as different plastics, and a larger amount of hazardous materials, such as lead [23].

Effective electronics recycling requires that consumers both have access to recycling programs and have knowledge of such programs effectiveness. This essentially means that consumers need to know where to take their electronic devices when they become obsolete or defunct. For some time now, municipalities have piloted collection programs that are available to the consumer [20]. As a result, ongoing electronics collection programs are more common. Their success requires advertising, ease of access, and awareness for consumers to understand the costs and benefits of recycling, especially for special event collection [22, 47]. Collection might involve either a permanent collection site or a special event site. Appropriate advertising methods should be selected according to regional characteristics. Methods differ between rural and urban areas, and residential and business areas. To date, collection efforts have overlooked the

Table 5.1 Summary of collection options and transportation responsibilities

Collection options	Transportation to collection site	Transportation to recycling site	Advantages	Disadvantages
Curb side	NA	Local government or recycler	Convenient residential participation	Potential for theft and abandonment, extra sorting, and high transportation costs
Special drop off event	Consumer	Local government or recycler	Increased recycling awareness, good for rural areas	Irregular collection amount, need storage space
Permanent drop off event	Consumer	Local government or recycler	High sorting rate, low transportation cost, and most cost effective	Need regular checking, not effective for all communities
Take back	NA	Original equipment manufacturer (OEM) or contract with OEM	No collection site needed	High shipment costs, need special packaging, and consumer visit to shipping location
Point of purchase	Consumer	Retailer	Low cost, high visibility if promoted by retailer	Retailer commitment and storage space

New collection options include the emerging industry of business to business collection. Opportunities exist to increase amount of CE and IT assets for all collection options if consumers know sensitive data will be destroyed properly before recycling

importance of quality assurance at the collection events and having consumers know sensitive data can be protected through transparent destruction before salvage and recovery of recyclable materials. See Table 5.1 for examples of collection options and responsibilities as outlined by [23]. With the existence of collection efforts on a local level, there have been several international and national initiatives aimed at improving recycling and reducing the impacts of e-waste.

5.3 Efforts to Reduce Environmental Impact and Promote Recycling

Major international and national efforts have been in place to limit illegal disposal of CEs, to promote reverse logistics and support CE recycling. A review of the predominant policies and efforts within the top global economies of the United States, Europe, and Japan reveals a consistent trend toward better management of CE waste and resources. These same trends include signaling to manufacturers about the importance of EPR programs. Differences include the focus of these efforts between shipping, product reclamation, reduction of hazardous materials within products, and the use of environmental management systems.

In 1989, the Basel Convention, the leading international authority on regulating the reduction of e-waste, was established to control "transboundary movements of hazardous wastes and their disposal" among countries [45]. Signatories to the Convention are not allowed to import or export any electronic components that may contain toxic chemicals [34]. Currently, there are 175 participants of the Basel Convention. Interestingly, Afghanistan, Haiti, and the United States have not yet ratified the document [2, 46].

Since the inception of the Basel Convention, European communities have presented a detailed document that specifies methods for regulating e-waste. The Directive on Waste Electrical and Electronic Equipment, or WEEE Directive, was presented and accepted by the European Parliament in 2003. The directive's purpose was to prevent e-waste from becoming a problem by reusing or recycling recoverable electrical parts [15]. An amendment to the original directive forced producers to internalize external costs, such as recycling and proper disposal, instead of burdening the consumer with costs for proper disposal [16].

In 2001, Japan implemented their Specified Home Appliance Recycling Law, which requires manufacturers to take back their electronic products and home appliances [31]. The national law also makes it illegal to dump any e-waste or home appliance in municipal landfills or roadsides [31]. In 2003, the Organization for Economic Co-operation and Development (OECD) completed the Environmentally Sound Management of Waste report, which advises countries on collecting, disposing, storing, and recovering their hazardous e-waste [36]. This report, however, can only make recommendations to waste facilities about environmental management systems; auditing environment, health and safety measures; and monitoring and recording emissions, and waste generation [36]. The report cannot force countries to implement these measures.

In North America, The Commission for Environmental Cooperation created a proposal in 2004 similar to the WEEE Directive that was introduced in Europe. The proposal was not well received by electronics-related industries [48]. Adoption of the proposal to the Commission did not take place [48] leaving the burden of developing formal e-waste initiatives on the shoulders of many proactive corporations and states. In March of 2010 the Basel Action Network (BAN), located in Seattle Washington, released the E-Stewards Standards for Responsible Recycling (R2) and Reuses of Electronics Equipment Program. Intended for international use, this program is consistent with international trade rules, social accountability standards, and aligned with environmental management systems standards. While e-steward certification is not a law calling for better management of electronics waste streams, it is a new form of quality assurance certification for supply chain members responsible for the management of CE and IT assets within a reverse logistics system. For a summary of international efforts impacting CE and IT asset recycling see Table 5.2.

Table 5.2 International efforts impacting CE recycling

Participating countries	Governing body	Initiative	Adoption
167 Countries of the UN (excluding Afghanistan, Haiti, and the US	United Nations Environmental Programme (UNEP)	Basel convention: no transboundary movement of hazardous waste	1999
Japan	Ministry of the Environment, Government of Japan	Home appliance recycling law: home appliance manufacturers must take back and recycle end use products	2001
The European Union	European Parliament	WEEE directive: reuse/recycle electronic parts; manufacturers internalize take back and recycling costs	2003
Intended for OECD Countries	Organization for Economic Co-operation and Development (OECD)	Environmentally sound management of waste: reclaim e-waste	2003
Written for International Use	Basel Action Network (BAN)	Industry specific environmental management system (EMS) standard designed as the basis for "E-Stewards" certification, consistent with international trade rules, social accountability standards, and EMS norms	2010

(Modified from [34])

The US stance on the Basel Convention has not stopped private organizations and local governments within US national boundaries from introducing local policies to regulate electronic equipment. Some corporations have played a pro-active role in the e-waste debate by providing electronic recycling programs: Dell, eMachines, Gateway, Hewlett-Packard, IBM, Lexmark, NEC, and Toshiba, for individual consumers, and Xerox for large-scale office equipment customers (SVTC 2004a, b). A snapshot of continued regulatory trends finds over 20 states with initiatives in place and as of the writing of this chapter, Pennsylvania has become the 24th state in the US to pass an EPR program and Landfill ban for TV's, Monitors, and Computers.

As the largest generators of e-waste in the world, major corporations have developed policies that promote responsible recycling. Yet, the vast majority of the IT assets and other electronics recovered from businesses in developed countries is exported to developing countries, where it is devastating the environment and human health. Despite international treaties and Europe's WEEE directive, e-waste export continues to flow freely. Why? Unethical recyclers promise to do the right thing— then increase their profits by shipping their customers' toxic-laden computers to Asia or Africa for dismantling under horrific conditions.

The E-Stewards Standards for Responsible Recycling and Reuse of Electronics Equipment Recycling Certification is the first fully accredited, independently audited certification program designed to ensure that e-waste will be responsibly—and accountably—recycled, and not dumped in developing countries. The E-Stewards Certification was developed by a group of concerned electronics recyclers, environmentalists, industry leaders, and health and safety and technical experts working with the BAN, a nongovernmental organization (NGO) focused on stopping illicit e-waste exports [1].

The exponential growth of e-waste combined with regulatory trends indicates the growing importance of understanding life cycle impacts of CEs. This process begins in new product development and extends to end use of CE products with improved opportunities for cradle-to-cradle thinking, environmental quality assurance, design for disassembly and recycling, and not disposal.

5.4 The Evolution of Environmental Focused Quality Assurance

Since the 1980s and the call to improve quality in the US, a large amount of research has been conducted regarding various, significant quality issues. Existing frameworks for quality and supply chain management stress the importance of relationships [28], communication [4], agility [27], speed, and supplier selection [5] to name a few. However, little research has focused on product and process quality assurance programs and the use of international standards for recycling of consumer electronics.

Quality management and more recently environmental management are important to the competitive capabilities of any organization. The importance of assuring product and process quality requires that it is not dealt with on an ad hoc basis. A properly implemented quality management system, and to go a step further, an Environmental Management System (EMS), within an organization and across its supply chain can provide protection from short-term actions which do not serve long-term goals. For many firms, obtaining acceptable levels of quality comes with the certification of a quality management system for itself and its suppliers. The international organization of standards (ISO) provides in ISO 9000:2008 what is regarded as the most prevalent approach to developing a quality management system. The continued growth of this standard for nearly 25 years suggests that it is, and will continue to be, an influential global standard [51, 25].

Studies show that many companies acquire ISO certification under pressure by their customers [12]. Lack of internal enthusiasm and internal motivation can hinder the impact of ISO 9000 on firm performance. If quality management is viewed as a process emphasizing management practices throughout the firm, then firms pressured into certification only for the sake of having certification may overlook the opportunity to change management practices and enhance quality performance. Yet, Corbett et al. [8] demonstrate significant improvements in financial performance for firms certified to the ISO 9000 standards. There appears to be an opportunity for firms within recycling supply chains to consider ISO 9000 certification to enhance customer and stakeholder perception, as well as quality assurance of the CE and IT assets recycled.

As an evolution of quality management, ISO 14001 was published in 1996 and provides the basic framework for the establishment of an EMS. The ISO defines an EMS as that part of the overall management system which includes organizational structure, planning activities, responsibilities, practices, procedures, processes, and resources for developing implementing, achieving, reviewing, and maintaining an environmental policy [6, 10]. These formal systems provide an opportunity to integrate reverse logistics systems and take back programs into planning, proce-dures, and processes that help to measure and manage CE assets throughout the product life cycle. The five requirements of ISO 14001 include [9, 11]:

- Formation of a corporate environmental policy and commitment to an EMS,
- Development of a plan for implementation,
- Implementation and operation of the EMS,
- Monitoring and possible corrective action,
- Top management review and continuous improvement

Basically, the firm must designate responsibility for achieving objectives and targets at each relevant function and level, provide the means for fulfilling the objectives and targets, and designate a time frame within which they will be achieved [40]. This same framework and planning will be beneficial to recycling firms and reverse supply chain members to better understand processes and objectives across functions.

The ISO 14000 series of standards are management tools and process standards. Note, it is not a performance standard and certification does not guarantee better quality or EMS performance [39]. In other words, these standards do not tell organizations what environmental performance they must achieve. Instead, the standards describe a system that will help an organization to achieve its own objectives. The assumption is that better environmental management will lead to improved environmental performance [11, 24, 52].

In designing the ISO 14001 specifications, the intent was to create a generic model that could be applied by any type and size of organization. The standards of the ISO 14000 family can be implemented within the whole organization or only in specific parts; they are not industry or sector specific. These system specifica-tions merely represent the minimal requirements for registration against the ISO 14001 standard [3].

How has quality assurance evolved to meet the current needs of reverse logistics systems? Given the long-standing interest in ISO guidelines and certifi-cation, there is a proven path for reverse logistics service providers and deman-ufacturers to leverage existing certifications. While the quality movement is not new, standards for quality have evolved since the 1980s and have been out in front of many companies looking for a structured approach to develop both quality, and since the 1990s, environmental management systems. This evolution now includes a focus on environmental quality assurance and the possibility for companies to demonstrate emerging forms of quality assurance within new and growing industries. Both practitioners and scholars will want to know; are there other certifications available that are more applicable to this growing industry and how are some demanufacturers better prepared to meet the contemporary needs of both end users of CE products and corporate IT assets? The answer to these questions and new insight can be found in a focused, case study approach examining emerging small businesses involved in demanufacturing.

5.5 Methods

Methods for this chapter include both secondary and primary data collection and synthesis. Secondary data collection is found throughout the chapter within a review of the literature involving of e-waste trends, supply chain management, reverse logistics, response to legislation, standards, extended producer responsibility, and the evolution of quality assurance. The literature review provides a foundation for emerging trends, opportunities for business practices, and a gap in the literature overlooking the role of small demanufacturers in recycling system development.

Primary data collection involves a single case study. This section of the chapter highlights the importance of case studies and theory development, before looking at a small, leading edge company, whose operations reveal the future direction of quality assurance within the CE recycling industry.

5.5.1 Case Study Methodology

In instances where a well developed set of theories regarding a branch of knowledge does not exist, Eisenhardt [14], and McCutheon and Meridith [29] suggest theory development can be best done through case study research. In this stage of theory building, a key objective within a reverse logistics and demanufacturing environment is to identify and characterize new types of quality assurance practices used for CE products. With the focus of this research being exploratory in nature, qualitative data collection methods were used.

Field-based data collection methods were used to ensure the identification of important relationships among products, processes, quality assurance, standards, and certifications. As demonstrated by Eisenhardt [14], field-based data collection methods also help develop an understanding of why measures and relationships are important. The development of an interview protocol, (see Appendix), was based on a review of the literature, and the researchers' general understanding of quality assurance issues facing the reverse logistics industry today. The protocol was pretested with a team of MBA students conducting research with a reverse logistics and recycling service provider and then utilized for the firm included in this study. Questions focused on

- Quality assurance issues,
- Third party designations and certifications,
- Supply chain management impacts, and
- Challenges involving demanufacturing that impact operational performance metrics such as cost, time, flexibility, and quality.

Interviews were conducted in the respondent's facilities with the owner and management team, with follow up phone conversations and e-mail correspondents when requesting clarification. Questions and subsequent discussions focused on: the consideration of quality assurance as an important part of the reverse logistics process, the factors affecting demanufacturing, metrics, and perceived industry opportunities.

There are some pitfalls to case study analysis as revealed by Eisenhardt [14], including lack of simplicity or narrow and idiosyncratic theories. A primary disadvantage of the case research approach is the difficulty in drawing deterministic inferences, and there are limitations in terms of the external validity of the study. However, the sample selected for qualitative research, such as in this study, was purposeful as suggested by Eisenhardt [14] and Miles and Huberman [30]. Next steps were to identify a small, but growing company recognized as an innovator and early adopter [32] in the implementation and use of recycling and quality assurance programs. Multiple site visits, and work with its management to better understand its operations and marketing, provide the foundation for the following information.

5.5.2 Case Study: eLoop

This case study information illustrates how one firm sees the evolution of quality assurance and trends to reduce environmental impact while promoting recycling as presenting hidden challenges and new opportunities for small firms in the recycling industry. The following information leverages the case study methodology to highlight one firm's approach to the development of new business capabilities when recycling e-waste. Consequently, the specific case study fills a gap in the literature, providing anecdotal evidence of the role of small firms and quality assurance within a growing industry.

eLoop handles EOL equipment by disassembling, recycling the hazardous materials with qualified downstream partners, and recovering commodity grade materials, such as metals, plastics, and glass, for use in another form in another industry. Firms such as this, also called demanufacturers, complete a closed-loop system focusing on the grave to cradle activities that are central to advancing an industry. At the same time, they assure their customers of a quality management process that protects data integrity, while providing services including data compliance and destruction; refurbishing, remarketing and resale of IT assets; and ethical and environmentally responsible electronic recycling processes with less impact on the environment:

- Landfill avoidance of 98 % of all material,
- No untested/nonworking electronic equipment is exported to developing countries,
- All reclaimed commodity grade materials are resold.

Services start with the development of a comprehensive plan that evaluates an organization's risk. Unlike many CE recyclers, eLoop certify and control the destruction of sensitive data (even going to the extent to provide a picture of a destroyed hard drive); and then provide a complete chain of custody for equipment from the time it leaves a customer location until it reaches its final destination. All memory containing devices, such as hard drives, tape backup media and smart phones pose additional risk when computers, servers, and cell phones, are retired from service. eLoop brings a quality assurance solution to organizations looking to trust one company that can control, protect, and destroy the critical data that resides in memory on these devices and handle the proper reuse or recycling of memory containing devices.

Since 2009, this firm has grown from four employees, to having 18 people across all functional areas of the organization, in two locations: the two drop off and tear down facilities are located outside large metropolitan areas in Pittsburgh and Lancaster, Pennsylvania. During that period of time annual revenues are up 290 %, with projected revenues of $1 M within another year. The unique aspect of this recycling company goes beyond the collection and recycling of CE components to the development of a business offering services in project management that can act as an extension of an organization's IT or Environmental Health and Safety (EH&S) departments to support projects related to the retirement of IT assets. While the demands of most IT departments are focused on support and development emerging opportunities for recycling, eLoop services include hardware maintenance, data security, asset disposition logistics, and EOL recycling.

Put yourself in the shoes of a consumer or manager wanting to recycle old computers and phones, and then contemplate how a recycling firm establishes credibility and quality assurance for you. eLoop is a demanufacturer for a number of different end user and corporate clients. This firm works with organizations of all sizes is to protect reputational and compliance risk. These same services are available to the community through private and publicly sponsored community collection events and five permanent collection sites. As a Department of Environmental Protection[1] permitted demanufacturing facility, eLoop provides landfill diversion and auditable environmental compliance with a completely transparent chain-of-custody. Customers find quality assurance in efficient, secure, and

---

[1] The DEP audits and provides permits for these types of facilities and they were among the first to get this permit.

sustainable solutions available for CE and IT assets backed by certifications positioning the concept of product and process quality assurance in a new way.

To date, this company is a BANE-Steward, they are a member of the Institute for Scrap Recycling Industries, and they are training a staff to be "certified secure destruction specialists" through the National Association for Information Destruction (NAID). These credentials assure the consumer that the recycling process is environmentally sound and the data destruction services are following best industry practices. It is through these relationships and certifications that a company can establish both credibility and quality assurance. When asked "in what ways do designations help ensure quality processing of specific components or materials?" Ned Eldridge, the President and CEO of eLoop claims "It is all about proving your processes are in compliance and safe. In addition designations help to provide a transparent chain of custody that can be followed from taking possession of the materials through the reuse, recycling, and recovery options for the materials. Third party designations create accountability which leads to trust in the marketplace."

Until a firm is tasked with measuring and managing quality within the reverse logistics process, small-scale operations may not believe they have quality issues. As firms involved in reverse logistics activities are asked to work through an ISO or other industry certification processes, quality issues will surface. eLoop is pursuing ISO 14001 registration and still weighing the benefits of ISO 9000 registration. Receiving approval for its EMS indicates that eLoop successfully meets or exceeds international standards in the areas of minimizing harmful effects on the environment caused by its activities, and has achieved evidence of continual improvement of its environmental performance. As firms such as this one experience growth within a region and across the state in adding operating facilities, benefits will be seen from this certification through higher levels of systems control and integration. Management of reverse supply chains will see change coming from organizations such as the E-Stewards, who are quickly moving toward a certification process that will require firms to formalize recycling industry operating standards which include a EH&S program, ISO 14001, OSHAS 18001, and a third-party audit of these systems. This certification process has taken almost a full year to complete in 2011.

When determining where in the demanufacturing process quality certification programs will have impact, the handling of monitors and TVs do not provide as much of an opportunity as this equipment can be passed along to others in the supply chain. Computer equipment and CPUs are where quality is more important and firms have an opportunity to generate more revenue from erasing hard drives and resale. Additionally, with the forthcoming legislation demanufacturers can imagine that large brand name original equipment manufacturers (OEMs) will only want to be involved with a reverse logistics system where all enterprises in the system will have a certified quality management program and some of the larger, dominant supply chain recyclers will be setting the rules for supply chain quality assurance. This will force transparency and better management of processes up and down the supply chain system.

Table 5.3 Quality assurance visibility

Challenges

Operational capabilities due to size and scale

Client's lack of understanding of compliance issues

Systems integration for supply chain collaboration

Education of public and corporate clients

Changing regulatory climate

Capabilities

Pledged E-Steward with the basal action network and third-party auditing

State permitted recycling facility

Institute for scrap recycling industries membership

Recycling industry operating standards

Chain of custody/certification of destruction process

Opportunities

ISO 14000 and 9000 certifications

Pending regulations regarding disposal bands for e-waste

Increased awareness of the hazards generated from e-waste

Increased awareness of sensitive data within CE and IT assets

NAID certified of secure destruction specialists

The synthesis above provides insight regarding a small demanufacturing firm's positioning for quality assurance visibility within a growth industry. Other SMEs along with large reverse supply chain members and OEMs can learn from small firms and better position their own processes for future transparency and quality assurance opportunities

eLoop has positioned itself as a first mover in the industry. Recent trends have pushed consumers and corporations to seek recycling companies that can provide a transparent chain of custody from the time EOL assets leave a facility, through the demanufacturing process, to the downstream vendors that trade the commodities. Other certifications that help customer awareness of quality assurance include compliance with the Recycling Industry Operating Standards (RIOS) as outlined by the Institute of Scrap Recycling Industries (ISRI) of which this firm is a member.

The case study of eLoop contributes to our understanding of regulatory compliance, the use of existing and emerging certifications, and the important role of protecting privacy of those who recycle their CE and IT assets, within the context of a small demanufacturer. Quality assurance has become one capability for some firms who are able to demonstrate regulatory compliance and data security through the use of certification of facilities and management systems. For a summary of the challenges and opportunities eLoop faces, see Table 5.3. This case study helps to highlight the growing importance of quality assurance opportunities for small firms that in many places are the foundation of the recycling system infrastructure.

5.6 Emerging Markets and Role of Quality Assurance

End users and businesses today are faced with a variety of risks associated with the disposal of CE and IT hardware. In an age of reliance upon electronic devices, the storage of electronic records to increase operational efficiency and enhance profitability has become increasingly important. Now, businesses must address compliance risks associated with the accumulation of sensitive data. The improper disposal of electronic devices containing sensitive data, such as credit card numbers, medical records, financial records, and social security numbers, can expose a business to financial, legal, and public relations risks. In a PC World article, Spring [41] mentions that he bought and salvaged used hard drives and found sensitive business and personal data on nine out of ten drives.

The US Government currently mandates that businesses follow regulatory compliance to protect sensitive data through measures such as:

- The Federal Information Security Management Act,
- Health Insurance Portability and Accountability Act,
- Sorbanes-Oxley Act,
- Fair and Accurate Credit Transaction Act, and
- The Gramm-Leach-Blieley Act.

In December 2009, the US Government passed H.R.[2] 2221, which requires "reasonable security policies and procedures to protect data containing personal information, and to provide for a nationwide notice in the event of a security breach." Recently, the state of Pennsylvania has proposed a bill that will address e-waste; H.B.[3] 708 requires producers to develop take back program for CE devices. The specific bill focuses primarily on the recovery of computer monitors, desktops, and notebook computers, but is only one bill in an increasing trend of international, federal, state and local legislation pushing supply chain transparency, use of new standards and certification, and increased attention to quality assurance within reverse logistics and recycling of CE and IT assets.

It is important to understand relationships between costs and quality for recycling businesses. Certification raises a barrier to entry for this industry that helps to ensure the e-waste exporters and small, back yard metal recyclers will not be able to participate in new mandated programs. The timing in Pennsylvania is right, as new legislation will support a state wide and private program for OEMs. Ned Eldridge claims "the OEMs want the work done in a timely manner and without mistakes to avoid the risk of any negative publicity. These same OEMs will be the ones to set the fee process for services provided for the certified recyclers in their supply chain." The state will set recycling goals for the OEMs based on market share in the state and per capita goals. Most states begin with a 3 lb/capita goal (36,000,000 lbs in Pennsylvania) for the first year.

The combination of all of the above will bring increased volume of e-waste from consumers into the waste stream that will be subsidized by the OEMs, in addition to existing collection options. Better consumer awareness and the push for OEMs to be involved will lead to better pricing for recyclers as they reuse, recycle, and recover the base materials. The importance of costs cannot be overlooked [17].

---

[2] House of Representatives.

[3] House Bill.

This industry will be a commodity driven business and OEMs will have to help pay for recycling and supply chain infrastructure. When commodity prices are high recyclers will focus on effectiveness of operations. If commodity prices are low, recyclers will need to focus on efficiency, and effectiveness of operations, but may still find a positive net present value in operations. In the future demanufacturers and those involved in reverse logistics may pass the cost of recycling to the OEMs and these costs will flow through established nationwide networks. Successful recycling/demanufacturing firms and reverse logistics providers will be those that can provide certified processes, transparency, and better quality assurance than their competition.

5.7 The Frontier of CE and IT Asset Recycling

How can an end user understand if a recycler is providing services with good quality management? In the last few years, businesses and consumers have come a long way in understanding the recycling industry. First they accepted someone taking their equipment off site for free or even paying the business a small amount for the materials. However, the tremendous impact of the 60 min documentary "Following the Trail of Toxic E-Waste" in 2009 and the real life situations in many cities, where recyclers have been exporting e-waste to other countries, has caught the attention of consumers and corporate IT asset managers. Now customers ask, "what are you doing with our equipment?" Rarely, did clients want proof of their CE assets being recycled or the data destroyed. In the past, clients just asked the question "will or has it been recycled?" Only in the past couple of years, have a business people asked to see a complete chain of custody and a record keeping system that shows material movements, certificates of recycling, and transparency downstream as to what the next processor does with the material. These trends stress the importance of reverse supply chain systems and collaboration leveraged by IT [21]. The new questions center around the network of demanufacturers being used in a firm's chain of custody and how large is the firm's carbon footprint when all of the transportation is taken into account.

What does the future of quality management look like in the CE recycling industry? The global perspective is clear; trends show transparency, and management of CE assets over their entire life cycle as growing in importance. There are more opportunities than ever for reliable reverse supply chains, economic development, quality assurance, and better asset management and recovery. Taking a more local perspective, insight comes as states of the US pass legislations promoting EPR programs and landfill bans for CE products. When programs, such as these, pass the recycling responsibility back onto the OEM, the challenge will be to position firms like eLoop as third-party certified and having the appropriate quality assurance designations so as to ensure trust from the OEMs to recycle their CE products. The certification process that eLoop is going through will qualify them to work with OEMs, as these manufacturers will not do business with recyclers that do not achieve either BAN certification or R2 certification. H.B. 708 mandates that recyclers need to be certified to then be included in the Pennsylvania state program.

The frontier of CE and IT asset recycling is not made transparent through the lens of one case study, but the insight from forward thinking companies, such as eLoop, positions new information as to the impacts of changing regulations and emerging markets. Insight from this case study shows the role small demanufacturers have in developing recycling system infrastructure and how entrepreneurs and supply chain professionals can leverage quality assurance to overcome challenges, and position for emerging market opportunities.

5.8 Conclusions

There is a dearth of information within the existing literature on the role of small demanufacturers within recycling systems and how these entities leverage emerging quality assurance standards and certifications. A primary contribution of our study is found in understanding how small firms are overcoming emerging challenges (i.e., scale, need to educate clients and public, supply chain systems integration, and a changing regulatory landscape) and taking advantage of opportunities (i.e., ISO certifications, pending regulatory change calling for extended producer responsibility, increased awareness of environmental hazards associated with disposal of e-waste, EOL data destruction, and new certifications for quality assurance), facing them within reverse supply chains focusing on the recycling of CEs and IT assets (see Table 5.3 for summary information of challenges, opportunities, and the relevant competitive capabilities). Moreover, the field study sheds new light on how quality assurance has evolved through the use of existing ISO standards and new certifications to meet the needs of reverse logistics and recycling systems. This understanding of environmentally focused quality assurance provides a means to uncover contemporary opportunities for emerging business models within the industry, and implications for the future of reverse logistics practices and research.

There is a lot to be optimistic about, as economies recover and new opportunities present themselves at international, national, and local levels. The optimism spills over to the ability for future research in emerging economies and recovering economies regarding what capabilities need to be present for the successful development of reverse logistics service providers, demanufacturers, and quality assurance of CE asset recycling. Researchers from all disciplines will have rich areas of inquiry involving the transitions of societies from a throw-away mentality to that of a waste as a resource, or cradle-to-cradle mentality. Scholars should look to emerging new product development standards that involve sustainability earlier in their processes, to identify the important relationships and impacts throughout supply chains. Other areas of research with the ability to leverage, this study include explaining and predicting the growth and eventual integration of recycling industries; impacts of transparency; regulatory change; consumer behavior; EPR; relationships between quality standards within the reverse logistics industry and performance; and the role of environmental management systems within manufacturing firms and the integration of supply chain members.

As the volume of goods coming into reverse logistics systems increases, protecting sensitive data, quality assurance and transparency throughout reverse logistic systems will be the norm. Whether motivated by governmental regulatory pressure, the potential for new sources of competitive advantage, quality assurance, or better environmental management, an increasing number of companies will consider the life cycle impacts of their products so that landfills are no longer a viable option. Organizations should approach designing reverse logistic systems in conjunction with quality assurance certifications realizing that early adoption of these certifications will become mandatory. Future networks and companies providing a bundle of services, which consider a cradle-to-cradle perspective by looking at both forward and reverse flows in their initial design, have the most potential for creating value and long-term competitive advantages for any organization.

Acknowledgments A special thanks to Preeti Srivastav, a MBA+Sustainability Fellow, for her research and help with this chapter and related projects.

A.1 Appendix. Interview Protocol

1. How many employees, size of the company?
2. What designations does your business have?

   a. Do any of these designations help ensure "quality management" of processes?

3. When or where in the supply chain do these designations help ensure quality management?
4. In what ways do the designations help ensure quality processing of specific components or materials
5. How can a business better understand if a recycler is providing services with good quality management?
6. Where do you experience quality issues in reverse logistics?
7. How can consumers and businesses better help ensure quality materials?
8. What tools and practices are available to businesses like eLoop to help ensure better quality?
9. What does the future of quality management look like in the CE recycling industry?
10. Are there any relationships between cost, quality, time, or flexibility when recycling CEs?
11. Would ISO 9000 or ISO 14000 certification be considered?

12. What part of the demanufacturing processes are best for having quality certification?

    a. Collection, test/sort, disassembly, reduction, separation by material, and commodity market

13. Are there any relationships between quality, cost, flexibility, and time in your industry?

14. What part of demanufacturing processes will be impacted by quality certifications?

References

1. Basel Action Network (BAN) (2011). http://www.ban.org/main/about_BAN.html
2. Basel Convention (2011). http://www.basel.int/ratif/convention.htm
3. Berthelot S, Coulmont M (2004) ISO 14000–a profitable investment? CMA Manage 78(7):36–39
4. Cai S, Jun M, Yang Z (2006) The impact of interorganizational internet communication on purchasing performance: a study of chinese manufacturing firms. J Suppl Chain Manage Summer 16–39
5. Choi T, Hartley J (1996) An exploration of supplier selection practices across the supply chain. J Oper Manage 14:333–343
6. Corbett C (2006) Global diffusion of ISO 9000 certification through supply chains. Manuf Ser Oper Manage 8(4):330–350
7. Council of Logistics Management (CLM) (1998) What it's all about. Council of Logistics Management, Oak Brook
8. Corbett CJ, Montes-Sancho MJ, Kirsch DA (2005) The financial impact of ISO 9000 certification in the United States: an empirical analysis. Manage Sci 51:1046–1059
9. Curkovic S, Sroufe R, Melnyk SA (2005) Identifying the factors which affect the decision to attain ISO 14001. J Energy 30(8):1387–1407
10. Darnall N (2006) Why firms mandate ISO 14001 certification. Bus Soc 45(3):354–382
11. Darnall N, Jolley GJ, Handfield R (2008) Environmental management systems and green supply chain management: complements for sustainability? Bus Strat Environ 17(1):30–45
12. Davis B (2004) One standard fits all. Prof Eng 17:43–45
13. Drake MJ, Ferguson ME (2008) Closed-loop supply chain management for global sustainability. In: Stoner JA, Wankel C (eds) Global sustainability initiatives: new models and new approaches. Chapter 9. pp 171–190
14. Eisenhardt K (1989) Building theories from case study research. Acad Manage Rev 14:532–550
15. European Parliament and Council of the European Union (European Parliament) (2003a) Directive 2002/95/EC of the European Parliament and of the Council of 27 January 2003 on the restriction of the use of certain hazardous substances in electrical and electronic equipment. Official J European Union, Legislation Series. No. 37, pp 19–23
16. European Parliament and Council of the European Union (European Parliament) (2003b) Directive 2003/108/EC of the European Parliament and of the Council of 8 December 2003 amending Directive 2002/96/EC on waste electrical and electronic equipment (WEEE). Official J European Union, Legislation Series. No. 345, pp 106–107
17. Fassoula E (2005) Reverse logistics as a means of reducing the cost of quality. Total Qual Manag 16(5):631–643
18. Fleischmann M (1997) Quantitative models for reverse logistics: a review. Eur J Oper Res 103:1–17

19. Garfinkel SL, Shelat A (2003) Remembrance of data passed: a study of disk sanitization practise. IEEE Secur Priv 1(1)
20. International Association of Electronics Recyclers (IAER) (2003) IAER electronics recycling industry report
21. Jayaraman V, Ross A, Agrawal A (2008) Role of information technology and collaboration in reverse logistics supply chains. Int J Logist: Res Appl 11(6):409–425
22. Jung LB, Bartel T (1998) An industry approach to consumer recycling: the San Jose project. In: Proceedings of IEEE international symposium on electronics and the environment p 36–41
23. Kang H, Schoenung J (2005) Electronic waste recycling: a review of the U.S infrastructure and technology options. Resour Conserv Recy 45:368–400
24. Kannan G, Haq A, Sasikumar P, Arunachalam S (2008) Analysis and selection of green suppliers using interpretative structural modelling and analytic hierarchy process. Int J Manage Decis Making 9(2):163–193
25. Kartha CP (2004) A comparison of ISO 9000:2000 quality system Standards, QS 9000, ISO/TS 16949 and Baldrige criteria. TQM Magazine 16:331–340
26. Krikke H, Blance I, van de Velde S (2004) Product modularity and the design of closed-loop supply chains. Calif Manage Rev 46:23–39
27. Lee H (2004) The triple A supply chain. Harv Bus Rev 82:102–112
28. Liker JK, Choi TY (2004) Building deep supplier relationships. Harvard Bus Rev 32:104–113
29. McCutcheon D, Meridith J (1993) Conducting case study research in operations management. J Oper Manage 11(3):239–256
30. Miles MB, Huberman M (1994) Qualitative data analysis. Sage Publications, Thousand Oaks
31. Ministry of the Environment, Government of Japan (2005) Recycling of specified kinds of home appliances at municipalities. Online at http://www.env.go.jp/en/press/2005/1027a.html
32. Moore G (1991) Crossing the chasm. Harper Business, New York
33. Nagurney A, Toyasaki F (2005) Reverse supply chain management and electronic waste recylcling: a mulitnetwork equilibrium framework for E-cycling. Transport Res E 41:1–28
34. Nakagawa L (2006) Toxic trade: the real cost of electronics waste exports from the United States. In: Cassara A, Damassa T (eds) Earth trends environmental essay competition
35. Northeast Recycling Council, Inc (NERC). Setting up and operating electronics recycling/reuse programs: a manual for municipalities and counties; March 2002. Cited within Kang, H. and Schoenung, J. 2005. Electronic waste recycling: a review of the U.S. infrastructure and technology options. Resour Conserv Recy 45:368–400
36. Organization for Economic Cooperation and Development (OECD) (2003) Working Group on Waste Prevention and recycling. Environmentally sound management (ESM) of waste (Output area 2.3.4) Project Fact Sheet
37. Parlikad A, McFarlane D, Fleisch E, Gross S (2003) Role of product identity in end-of-life decision making, White Paper, Auto-ID Center, Institute for Manufacturing, University of Cambridge, UK
38. Plambeck E, Wang Q (2009) Effects of E-waste regulation on new product introduction. Manage Sci 55(3):333–347
39. Rondinelli D, Vastag G (2000) Panacea, common sense, or just a label? The value of environmental management systems. Eur Manage J 18(5):499–510
40. Rowland-Jones R, Pryde M, Cresser M (2005) An evaluation of current environmental management systems as indicators of environmental performance. Manage Environ Qual 16(3):211–219
41. Spring T (2003) Hard drives exposed. PC World. Accessed June of 2011. http://www.pcworld.com/article/110012/hard_drives_exposed.html
42. Silicon Valley Toxics Coalition (SVTC) (2004a) Fifth Annual Computer Report Card
43. Silicon Valley Toxics Coalition (SVTC) (2004b) Poison PCs and Toxic TVs: E-waste Tsunami to Roll Across the US: Are We Prepared?. San Jose, SVTC

44. Toffel MW (2003) The growing strategic importance of end-of-life product management. Calif Manage Rev 45(3):102–129
45. United Nations Environment Programme, (UNEP) Secretariat of the Basel Convention (1989) Basel convention on the control of transboundary movements of hazardous wastes and their disposal
46. United Nations Environment Programme, (UNEP) Secretariat of the Basel Convention (2006) Parties to the basel convention. Online at http://www.basel.int/ratif/frsetmain.php. Accessed. Feb 2006
47. United States Environmental Protection Agency (USEPA) (2000) Electronic reuse and recycling infrastructure development in Massachusetts, EPA–901-R-00-002
48. United States Environmental Protection Agency (USEPA) (2004) Office of the Inspector General. Multiple Actions Taken to Address Electronic Waste, But EPA Needs to Provide Clear National Direction, Report No. 2004-P-00028
49. United States Environmental Protection Agency (USEPA) (2008a) Fact sheet: management of electronic waste in the United States. EPA530-F-08-014
50. United States Environmental Protection Agency (USEPA) (2008b) Municipal solid waste generation, recycling and disposal in the United States, facts and figures for 2008
51. Uzumeri MV (1997) ISO 9000 and other management meta standards: principles for management practice. Acad Manage Execut 21–36
52. Viadiu F, Saizarbitoria I (2006) ISO 9000 and ISO 14000 standards: an international diffusion model. Int J Oper Prod Manage 26(1/2):141–165
53. Zuidwijk R, Krikke H (2008) Strategic response to EEE returns: product eco-design or new recovery process? Eur J Oper Res 191:1206–1222
54. Zoeteman B, Krikke H, Venselaar J (2010) Handing WEEE waste flows: on the effectiveness of producer responsibility in a globalizing world. Int J Adv Manuf Technol 47:415–436

Chapter 6
Quality Assurance in Remanufacturing with Sensor Embedded Products

Onder Ondemir and Surendra M. Gupta

Abstract Emerging information technologies, such as sensors and radio frequency identification (RFID) tags could be used to mitigate planning of remanufacturing operations by reducing or almost eliminating uncertainty. Using the information collected by sensors, existence, types, conditions, and remaining lives of components in an end-of-life product (EOLP) can be determined. Remaining useful life can be taken into account as a good measure of quality. Therefore, determination of remaining useful life allows decision makers to construct sophisticated recovery models that guarantee a minimum quality level on recovered products while optimizing various system criteria. In this paper, we present a remanufacturing-to-order (RTO) system for end-of-life sensor embedded products (SEPs). An integer programming (IP) model is proposed to determine how to process each and every end-of-life product on hand to meet the quality-based product and component demands as well as recycled material demand while fulfilling the minimum cost objective. Demands are met by disassembly, remanufacturing, and recycling operations. Outside component procurement option is used to eliminate the component and material backorders. A case example is considered to illustrate the application of the proposed methodology.

O. Ondemir
Department of Industrial Engineering, Yildiz Technical University,
Barbaros Bulvari, Besiktas, 34349 Istanbul, Turkey

S. M. Gupta (✉)
Laboratory for Responsible Manufacturing, 334 SN, Department of Mechanical and Industrial Engineering, Northeastern University, 360 Huntington Avenue, Boston, MA 02120, USA
e-mail: gupta@neu.edu

Y. Nikolaidis (ed.), *Quality Management in Reverse Logistics*,
DOI: 10.1007/978-1-4471-4537-0_6, © Springer-Verlag London 2013

6.1 Introduction

Alarming increase in the use of natural resources and decreasing number of landfills have caused many environmental problems which have led to several government regulations that hold manufacturers responsible for their products after the products reach their end-of-lives (EOL). This has given rise to a phenomenon known as reverse logistics (RL), which involves collection, transportation, and management of end-of-life products (EOLPs). There are many advantages in managing EOLPs such as reduction in the use of virgin resources, landfill conservation, and cost savings stemming from the recovery of EOLPs.

Management of EOLPs consists of a series of operations such as cleaning, disassembly, sorting, inspecting, and recovery or disposal. Recovery options include remanufacturing, refurbishing, repairing, component recovery, and material recovery (via recycling). Quality of collected EOLPs plays a big role in determining the recovery option to choose. However, neither the quality nor the quantity of returning EOLPs is predictable. Hence, the outcome of the recovery operations is highly uncertain. This uncertainty is what makes quality management a challenging task in an RL setting. As one of the key elements of RL, remanufacturing exhibits, by far, the most difficult operations management problems. This is mostly because the variability and uncertainty associated with the quality of returned products lead to a huge variation in the product recovery operations and the quality of harvested components, spare parts, and remanufactured products. Lack of information on the components 'quality status necessitates comprehensive' testing and inspection. Remanufacturing operations to be performed and necessary spare parts to be used are decided based on the testing results. Each EOLP has its own quality condition and exhibits unique remanufacturing requirements because they originate from various sources where they had been subjected to different working conditions, usage patterns, and user upgrades. As a result, finding the EOLPs with minimal recovery costs requires testing the whole EOLP inventory. It should be noted that, for each product that turns out to be nonreusable, all the time and effort put in for inspection is wasted. In order to address this, companies try to influence the quality of returns by offering financial incentives to product holders to return their higher quality products [1]. On the other hand, incentives cannot guarantee the acquisition of sufficient EOLPs in the desired quality status. Preliminary disassembly and quality tests are still necessary. There are a number of studies in the literature [1–3] that assign a quality level to each EOLP based on the results of the inspection. In many cases, there is a finite discrete set of quality levels that are defined based on nonquantifiable data. Therefore, quality determination accuracy is highly unconvincing. This leads to financial losses in remanufacturing companies [4]. Embedded sensors have a potential to address aforementioned issues and mitigate quality assurance of remanufactured products by providing crucial life cycle information and enabling accurate quality determination.

A sensor is a monitoring device that keeps a log of the changes in the value of various measures such as temperature, pressure, and vibration. Sensor embedded products (SEPs) are built with sensors implanted in them to monitor their critical components while they are in use. Using the information collected by sensors, existence, types, conditions, and remaining lives of components in an EOLP can be determined. Remaining useful life is a good measure of quality. Once the remaining useful lives of components are known, quality-based customer demands can be met optimally. Although there are many types of sensors, active radio frequency identification tags (RFID) are the most suitable ones. Active RFID tags are capable of keeping static data (production date, sale price, sale date, etc.) and collecting dynamic life cycle data (run cycles, working temperature, shocks, vibrations, etc.) [5]. Data delivery is performed over radio waves without a line of sight. RFID tags deliver all the data to the central information system when the embedded EOLPs and the RFID readers are within a certain distance.

In this chapter, a remanufacture-to-order (RTO) system for EOL SEPs is presented. An integer programming (IP) model is proposed to determine how to process each and every EOLP on hand to meet the quality-based product and component demands as well as recycled material demand while fulfilling the minimum cost objective. Demands are met by disassembly, remanufacturing, and recycling operations. Outside component procurement option is used to eliminate the component and material backorders. A case example is considered to illustrate the application of the proposed methodology.

6.2 Literature Review

6.2.1 Product Recovery

Environmentally conscious manufacturing and product recovery (ECMPRO) through 1999 is presented by Gungor and Gupta [6]. Ilgin and Gupta [7] extended this work by reviewing more than 500 references published through 2009. Disassembly is one of the hottest research areas in ECMPRO owing to its importance in all recovery operations. Of all disassembly problems, the disassembly-to-order (DTO) literature includes the most relevant studies to this work. The goal of DTO is to determine the optimum number of EOLPs to be disassembled in order to fulfill the demand for components and materials such that some desired criteria of the system (cost minimization, profit maximization, etc.) are satisfied. Kongar and Gupta [8] presented a preemptive goal programming (PGP) model of an electronic products DTO system considering a variety of physical, financial, and environmental constraints and goals. In [9], the same authors proposed a linear physical programming (LPP) model to solve the DTO problem with multiple physical targets. Heuristics, meta-heuristics, and expert models have been used in the DTO problem because of the increased complexity especially when multiple criteria and/or periods are considered. Kongar and Gupta [10] presented a multi-objective TS algorithm and Gupta et al. [11] proposed an artificial neural network model in order to solve the DTO problem. Langella [12] developed a multi-period heuristic considering holding costs and external procurement of items. Inderfurth and Langella [13] developed two heuristic procedures to investigate the effect of stochastic yields on the DTO system. Kongar and Gupta [14] presented a fuzzy goal programming model for DTO systems under uncertainty. For additional reading on disassembly operations and problems, the reader is referred to [15–20].

6.2.2 Quality Issues in Product Recovery

Quality assurance is one of the main challenges encountered in product recovery. Necessary recovery operations and quality of the recovered products are highly dependent on the quality status of the returned EOLPs. A recent book by Ilgin and Gupta [21] covers quality assurance and house of quality in a remanufacturing setting. Behret and Korugan [22] analyzed the effects of uncertainties in return quality in a hybrid (viz., remanufacturing and manufacturing) system via simulation. The authors concluded that as the return rate increases, remanufacturing operations dominate the system and the quality-based classification of returns becomes much more important. van Wassenhove and Zikopoulos [4] investigated the effects of quality overestimation in an RL setting where the returned products are graded and classified based on a list of nominal quality metrics provided by the remanufacturer. Das and Chowdhury [23] proposed a mixed integer programming (MIP) model, which considers modular product designs and integrates an RL process for the collection of returned products, their recovery processes, and production of products at different quality levels in order to maximize overall supply chain profit. Again, the quality levels are defined by indirect terms that do not necessarily measure quality directly. Nenes and Nikolaidis [3] proposed a multi-period MIP model for the optimization of procurement, remanufacturing, stocking, and salvaging decisions considering multiple suppliers and several quality levels of returned products. Nikolaidis [1] investigated the profitability of remanufacturing used products using an MIP model assuming the returns might be in several different quality conditions. Chouinard et al. [2] proposed a stochastic programming model for designing supply loops considering five return product quality states (i.e., unknown, new, good condition, deteriorated or damaged, and failing). Denizel et al. [24] presented a stochastic programming approach considering certain probabilistic quality scenarios. Pochampally and Gupta [25] implemented the six sigma quality approach for the selection of potential recovery facilities in reverse supply chains. Kim [26] introduced the quality embedded remanufacturing (QRS) and proposed a multi-agent approach and a real-time scheduling mechanism for the QRS along with modeling tools and quality embedded dispatching rules. El Saadany [27] developed two inventory models with quality considerations for hybrid systems where manufacturing and remanufacturing options were utilized. Model I considers the production and remanufacturing with imperfect quality considerations, and Model II presents reworking as a method to correct defectives, in a RL inventory context. Chiu-Wei [28] focused on the cost factors and environmental performance associated with Cost of Quality and environmentally friendly manufacturing.

6.2.3 Remote Product Life Cycle Monitoring Devices

With the advent of Internet, wireless communication, and product identification technologies, product life cycle data collection has gained attention in recent years. Having visibility of product information over the whole life cycle allows for better decision-making. The core element of these technologies is the product embedded information device (PEID) technology [29]. PEIDs may be a combination of active and passive RFID tags, and sensors. Active and semi-active RFID tags are capable of collecting environmental data [5]. Schmidt and van Laerhoven [30] proposed the use of embedded sensors to build environment discerning products to enhance user interfaces, and facilitate communication during the middle of life (MOL). Soroor et al. [31] proposed mobile real-time supply chain coordination (MRSCC) approach that utilizes the data collected by sensors. Cheng et al. [32] developed a generic embedded device (GED) that can be attached to equipment, portal servers, and automatic guided vehicles to retrieve, collect, and manage equipment data. In particular, the communication management module of GED can transmit and receive data to/from remote clients via both wired and wireless networks. Chang and Hung [33] developed a new measurement technique that enables direct, real-time measurements, and continuous monitoring of concrete internal temperature and humidity for maintenance purposes. The technique makes use of radio frequency integrated circuits and humidity/temperature sensors. Meyer et al. [34] presented an overview of the intelligent products and discussed their potential use in manufacturing, supply chain, and product life cycle management areas. A number of methods were developed to collect life cycle data in various phases of supply chains [35–37].

Klausner et al. [38] proposed an information system for product recovery (ISPR) where a sensor is integrated in a product to record and store data strongly correlated with the degradation of components during the use stage of a product. The data recorded and processed during the use stage are retrieved and analyzed by the ISPR when the product is returned. Klausner et al. [39], extended the previous study by including the economic efficiency of the sensor in a reuse scenario. Yang et al. [40] performed field trials to show that SEPs can enable EOL treatment and other suitable product-related services. Vadde et al. [41] showed the use of sensor-based life cycle data in EOL decisions via a discrete-event simulation study. Ilgin and Gupta [42] investigated the impact of SEPs on the various performance measures of a multi-product (viz., refrigerators and washing machines) disassembly line using simulation analysis. The authors showed that SEPs not only reduce the total system cost, but also increase the revenue and profit. Also in [43], the authors simulated the recovery of sensor embedded washing machines using a multi-kanban controlled disassembly line. Similar studies and analyses can be found in [44, 45].

Waste electrical and electronic equipment (WEEE) directive is one of the most successful environmental regulations in the world. With WEEE, it has become mandatory to include the disassembly instructions with products [41]. RFID tags become prominent in environmentally conscious product recovery, because of its sufficient memory capacity to store disassembly and recycling information [46]. RFID technology will also be investigated, designed, and tested in the ZeroWIN (Towards Zero Waste in Industrial Networks) project [47]. Parlikad and McFarlane [48] discussed how RFID-based product identification technologies can be employed to provide the necessary information and showed the positive impacts on product recovery decisions. Gonnuru [49] proposed an RFID integrated fuzzy-based disassembly planning and sequencing model and showed the use of life cycle information for optimal disassembly decisions. Kulkarni et al. [50] examined the benefits of information provided by RFID tagged monitoring systems in decision-making during product recovery. Zhou et al. [51] proposed an RFID-based remote monitoring system for enterprise internal production management.

As mentioned above, the remaining life time of returned products is a good measure of quality and can be determined using sensor-based product life cycle data [52–57]. Herzog et al. [58] proved the advantage of using condition-based data in remaining life prediction. Mazhar et al. [59] proposed a two-step approach combining Weibull analysis and artificial neural networks (ANNs) for remaining life estimation of used components in consumer products. Byington et al. [60] proposed a data-driven neural network methodology to remaining life predictions for aircraft actuator components.

6.3 Remanufacturing-to-Order System

As opposed to a DTO system where traditional EOLPs are disassembled to harvest reusable parts and components, proposed RTO system includes remanufacturing, disassembly, and recycling of SEPs in order to meet quality-based product and component demands as well as material demands. Backorders are not allowed. Demand shortages are filled using the outside procurement option. Disassembly can be performed in a destructive or nondestructive manner. Although destructive disassembly is highly-cost effective, it renders the harvested parts nonfunctional. A sketch of the RTO system flow is given in Fig. 6.1.

In the RTO system, quality of a returned EOLP is defined based on the remaining useful life of its components which can be determined using the life cycle data collected by sensors. Each and every component in the EOLP is classified under a finite set of quality levels. For example, quality level "low" may refer to the items having at least 1 year of remaining life, similarly, quality level "high" may refer to the items having at least 5 years of remaining life. In order to maintain a good quality standard, items having less than a certain remaining life are considered to be "extremely low quality" and treated as nonfunctional components. Quantitative quality levels increase the accuracy and makes specific quality requirements attainable. Other advantages of SEPs over traditional EOLPs are summarized briefly in Table 6.1.

When EOLPs arrive at the recovery facility, all data stored on them are retrieved automatically by RFID readers. These data are used to determine non-functional and missing (removed) items, and the remaining lives (quality levels) of all functional items in the inventory. Then, based on their quality status and the

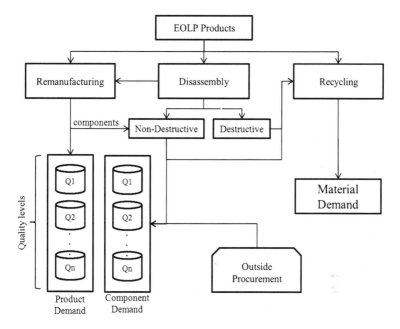

Fig. 6.1   RTO system flow

Table 6.1   Advantages to SEPs over traditional products in product recovery

Traditional products	SEPs
Disassembly is necessary in order to determine the status of each component	Functional, nonfunctional, and missing components in all EOLPs can be identified without preliminary disassembly
Component yield of an EOLP is unknown before disassembly	Component yield of each single EOLP is known when the data are retrieved
Quality is unknown before disassembly and inspection	Quality of each and every component in the inventory is known without disassembly and inspection
Optimization models use yield estimations derived from probability distributions or heuristic approaches	A deterministic model with actual data can be constructed. Solution to the model will be a detailed recovery plan
Quality-based demands can be considered with non-quantifiable vague terms	Quality-based demands can be considered and met with high accuracy
All components must be disassembled nondestructively no matter what the component conditions are	Disassembly operation is performed destructively or nondestructively based on the quality of the components in question

demands, EOLPs are remanufactured for product recovery or disassembled for component recovery. Environmental regulations impose a minimum acceptable recycling rate on nonreusable components (i.e., nonfunctional and extremely low quality components) regardless of actual material demand. It should be noted that the recycling cost includes the removal cost of RFID tags and sensors.

Remanufacturing option requires disassembly of nonfunctional components and may involve disassembly of the quality deficient functional components that do not meet quality requirements. Quality deficiency is defined based on the target quality level of remanufactured products. For instance, a low quality component is considered to be quality deficient when remanufacturing a high quality product, and needs to be replaced. Another situation where functional components may be disassembled is when there are multiple product types. Although products have common components, some components may be exclusive to certain product types. Remanufacturing operations may change the type of an EOLP. This may lead to the disassembly of extra components that are not in the bill of materials of the target remanufactured product type.

6.4 Mathematical Model

6.4.1 Nomenclature

Variables	Definition
$\bar{x}_i$	1 if EOLP i is disassembled nondestructively, zero otherwise;
xd_i	1 if EOLP i is disassembled destructively, zero otherwise;
$\bar{y}_i$	1 if EOLP i is remanufactured, zero otherwise;
x_{ijb}	1 if component j of EOLP i is disassembled and evaluated as b quality, zero otherwise;
y_{itm}	1 if EOLP i is remanufactured to produce a product t which will be evaluated as quality level m, zero otherwise;
rm_{ijb}	1 if component j of EOLP i is disassembled during remanufacturing and used as a b quality component, zero otherwise;
r_{jb}	Number of operable b quality component js that are recycled;
rb_j	Number of nonreusable component js that are recycled;
rep_{itmjb}	1 if a b quality component j is used to repair EOLP i to produce a product of type t for quality level m, zero otherwise;
l_{jb}	Number of b quality component js procured from outside,

Parameters	Definition
b, i, j, k, m, t	Running numbers;
I	Set of EOLPs on hand;
B	Set of quality;
J	Set of components dealt with;
M	Alias for B;
K	Set of material types dealt with;
T	Set of product types dealt with;
α	Minimum acceptable rate of recycling;
a_{ij}	1 if component j of EOLP i is available and functional, 0 otherwise;
f_{ij}	1 if component j in EOLP i is nonfunctional, 0 otherwise;
el_{ij}	1 if component j is available and functional in EOLP i, 0 otherwise;

(continued)

(continued)

Parameters	Definition
c_{jb}	Outside procurement cost of a b quality component j;
cin_{ijb}	1 if component j can be used as b quality, 0 otherwise;
cd_j	Disassembly cost of component j;
cr_j	Recycling cost of component j;
cb	Disassembly cost of a non-reusable component;
ca_j	Assembly cost of component j;
dc_{jb}	Demand for b quality component js;
def_{itmj}	1 if component j of EOLP i is quality deficient to produce a m quality product of type t, zero otherwise;
dm_k	Demand for material k;
dp_{tm}	Demand for m quality products of type t;
ext_{itj}	1 if component j of EOLP i is available but unnecessary to produce a product of type t, zero otherwise;
γ_{jk}	Material k yield of component j;
mis_{itj}	1 if component j of EOLP i is missing but necessary to produce a product of type t, zero otherwise;
rem_{ij}	Remaining life of component j of EOLP i;

6.4.2 Objective Function

The objective of RTO is to minimize the total cost (TC). TC can be written as the sum of total disassembly cost (TDC), total remanufacturing cost ($TRMC$), total recycling cost (TRC), and total outside procurement cost ($TOPC$). Therefore

$$TC = TDC + TRMC + TRC + TOPC \tag{6.1}$$

Each term is described below.

TDC is the cost of extracting all functional and nonfunctional components from all EOLPs that are to be disassembled. TDC can be formulated as:

$$TDC = \sum_{i \in I} \left[\bar{x}_i \sum_{j \in J} \left((1 - el_{ij}) a_{ij} cd_j + f_{ij} cb \right) + x d_i cb \sum_{j \in J} \left(a_{ij} + el_{ij} \right) \right] \tag{6.2}$$

$TRMC$ is incurred by the disassembly of nonfunctional, extra, and quality deficient components, and assembly of required ones. Therefore, $TRMC$ can be written as follows:

$$TRMC = \sum_{i,j \in J} \left[\bar{y}_i cb \left(f_{ij} + el_{ij} \right) + cd_j \sum_{b \in B} \left(rm_{ijb} \right) + ca_j \sum_{t \in T, m \in M} y_{itm} \left(\mathrm{def}_{itmj} + \mathrm{mis}_{itj} \right) \right] \tag{6.3}$$

TRC is defined as the cost of recycling nonreusable and reusable components. Assuming components may have different unit recycling costs, *TRC* can be formulated as seen below:

$$TRC = cr_j \sum_{j \in J} \left(rb_j + \sum_{b \in B} r_{jb} \right) \tag{6.4}$$

TOPC is a function of quality-based unit purchase cost and the total number of procured component *js*. Mathematical expression for *TOPC* can be written as follows:

$$TOPC = \sum_{j \in J, b \in B} c_{jb} l_{jb} \tag{6.5}$$

6.4.3 Constraints

An EOLP cannot be disassembled for its parts and remanufactured at the same time. Thus,

$$\bar{x}_i + xd_i + \bar{y}_i \leq 1, \qquad \forall i \tag{6.6}$$

Complete disassembly is imposed using the following constraints. These constraints also ensure that a component can be considered belonging to only one quality level after disassembly.

$$\sum_{b \in B} x_{ijb} = (1 - el_{ij}) a_{ij} \bar{x}_i, \quad \forall i, j \tag{6.7}$$

$$\sum_{\{b \in B | cin_{ijb} = 0\}} x_{ijb} = 0, \quad \forall i, j, b \tag{6.8}$$

Equations below ensure that, if chosen, an EOLP is remanufactured to produce only one product and quality-based demands are satisfied by these remanufactured EOLPs.

$$\sum_{t \in T, m \in M} y_{itm} = \bar{y}_i, \quad \forall i \tag{6.9}$$

$$\sum_{i \in I} y_{itm} = dp_{tm}, \quad \forall t, m \tag{6.10}$$

All reusable components harvested by disassembly and remanufacturing operations are collected and used to satisfy the internal and external component demands with respect to their quality levels. In case these components are not enough to meet the demands, outside procurement option is used to prevent backorders. Therefore,

$$\sum_{i\in I}\left[a_{ij}x_{ijb} + rm_{ijb} - \sum_{t\in T, m\in M} rep_{itmjb}\right] + l_{jb} - r_{jb}jb, \quad \forall j, b \tag{6.11}$$

where,

$$\sum_{\left\{b\in B|cin_{ijb}=0\right\}} rm_{ijb} = (1 - el_{ij})\sum_{t,m\in M} y_{itm}(def_{itmj} + ext_{itj}), \quad \forall i, j \tag{6.12}$$

$$\sum_{\left\{b\in B|cin_{ijb}=0\right\}} rm_{ijb} = 0, \quad \forall j, b \tag{6.13}$$

Missing and quality deficient components must be replaced with components of appropriate quality levels. Therefore,

$$\sum_{\left\{b\in B|b\geq m\right\}} rep_{itmjb} = y_{itm}(mis_{itj} + def_{itmj}), \quad \forall i, j, t, m \tag{6.14}$$

$$\sum_{\left\{b\in B|b < m\right\}} rep_{itmjb} = 0, \quad \forall i, j, t, m \tag{6.15}$$

Material demand is met by recycling components. These components may be reusable or nonreusable. It is allowed to recycle more material than the projected material demand in order to accommodate the minimum acceptable recycling rate regulation.

$$\sum_{j\in J}\gamma_{jk}\left(rb_j + \sum_{b\in B} r_{jb}\right) \geq dm_k, \quad \forall k \tag{6.16}$$

The total number of recycled nonreusable components (rb_j) has to be greater than a given percentage (α) of all nonreusable components according to the minimum acceptable recycling rate regulation. Obviously, total number of non-reusable components set the upper bound for rb_j. Therefore,

$$\alpha\sum_{i\in I}(f_{ij} + el_{ij})(\bar{x}_i + \bar{y}_i) \leq rb_j \leq \sum_{i\in I}(f_{ij} + el_{ij})(\bar{x}_i + \bar{y}_i), \quad \forall j \tag{6.17}$$

6.5 Numerical Example

This section illustrates the implementation of the above methodology to a case example that is similar to the one in [23]. The case is about a recovery facility that deals with 10 products. Each product consists of 6–10 components and configurations are given in Table 6.2. Except for the product types 6 and 7, all product configurations are adapted from the design 1 case of the aforementioned paper.

The recovery facility is equipped with RFID readers and accepts sensor and RFID embedded EOLPs. Hence, item level product life cycle data is acquired and used to determine the remaining lives of all components as soon as the products

Table 6.2 Configuration of the product types

Product Types	Components									
	1	2	3	4	5	6	7	8	9	10
1	☑	☑	☑	☑	☑	☑	☑	☑		
2	☑	☑	☑	☑				☑	☑	☑
3			☑	☑	☑	☑	☑	☑	☑	☑
4	☑	☑	☑	☑	☑	☑	☑	☑	☑	☑
5	☑	☑	☑	☑					☑	☑
6	☑	☑		☑	☑	☑	☑	☑	☑	☑
7		☑	☑	☑	☑	☑	☑	☑	☑	☑
8	☑	☑	☑				☑	☑	☑	☑
9	☑	☑	☑	☑			☑	☑	☑	☑
10	☑		☑	☑		☑	☑	☑	☑	☑

Table 6.3 Remaining useful lives of all components

EOLP ID#	Components									
	1	2	3	4	5	6	7	8	9	10
1	4.35	NF	5.05	3.45	–	–	–	3.29	3.11	5.96
2	4.67	2.86	–	5.03	1.31	6.57	NF	3.38	NF	6.58
3	3.03	4.30	2.63	7.58	–	–	–	–	NF	NF
4	3.83	4.45	4.21	NF	–	–	–	3.02	5.27	2.66
.	.	.	.	.	.	.	.	.	.	.
.	.	.	.	.	.	.	.	.	.	.
.	.	.	.	.	.	.	.	.	.	.
396	3.64	4.60	0.22	5.25	5.06	3.45	NF	–	5.73	3.23
397	6.54	7.33	NF	NF	–	–	–	–	4.36	5.61
398	4.01	3.63	3.92	–	–	–	NF	5.98	6.53	4.97
399	3.07	3.58	4.74	NF	–	–	4.88	4.07	1.75	4.62
400	3.15	3.99	6.01	2.43	–	–	3.64	5.27	NF	4.43

reach the facility. A small sample (9 of 400 rows) of the product information is given in Table 6.3. In this table, the numbers represent the remaining lives of corresponding components in years. "NF" means nonfunctional component, while "-" indicates nonexistent (missing) component. Based on this information, quality levels of components are identified.

Demands are assumed to occur for the items in three quality levels, namely, "low", "medium", and "high" quality items. Low, medium, and high quality products and components must have at least 1, 3, and 5 years of remaining lives, respectively. Items having less than 1 year of remaining life are considered to be "extremely low quality" and treated as nonfunctional. Two types of materials (materials A and B) are recovered using recycling. Quality-based demands, procurement costs, operational (disassembly, assembly, and recycling) costs, and material yields are exhibited in Table 6.4. Quality-based demands for remanufactured products are given in Table 6.5.

Table 6.4 Components demands, procurement and operational costs, and material yields

| | | Components | | | | | | | | | |
	Quality	1	2	3	4	5	6	7	8	9	10
Demand	Low	20	16	22	31	15	26	35	14	26	29
	Medium	22	12	22	20	31	33	33	17	23	25
	High	17	15	8	14	16	19	15	14	17	12
Procurement cost ($)	Low	20	40	60	15	15	15	12	18	22	24
	Medium	40	30	70	20	20	20	18	22	24	30
	High	50	60	75	25	25	25	24	26	30	32
Disassembly cost ($)	–		1.5	1.5	1.5	0.5	0.5	0.5	2	2	2	2
Assembly cost ($)	–		1.5	1.5	1.5	0.5	0.5	0.5	2	2	2	2
Recycling cost ($)	–		0.1	0.1	0.1	0.1	0.1	0.1	0.2	0.2	0.2	0.2
Material A yield (lbs.)	–		1	1	1	0	0	0	1.5	1.5	1.5	1.5
Material B yield (lbs.)	–		0	0	0	1.5	1.5	1.5	0	0	0	0

Table 6.5 Quality-based remanufactured product demands

| | Quality Levels | | |
Product types	Low	Medium	High
1	6	4	8
2	6	4	10
3	12	10	8
4	12	2	8
5	10	0	6
6	2	6	8
7	10	8	10
8	6	10	12
9	6	6	8
10	4	10	8

Demands for materials A and B are 600 and 300 lbs., respectively. Additionally, destructive disassembly cost is $0.20 ($cb = 0.2$),$B = M = \{$Low, Medium, High$\}$, $k = \{A, B\}$, $I = \{1, 2, \ldots, 400\}$, $J = \{1, 2, \ldots, 10\}$, and $\alpha = 0.7$.

The problem is modeled as an integer mathematical program using LINGO's programming language and solved by CPLEX Interactive Optimizer.

6.6 Results

CPLEX Interactive Optimizer is able to find the optimal solution to the model. The total cost of all recovery process is $5892.40. ID numbers of the EOLPs that are disassembled (destructively and nondestructively) and remanufactured in the optimal solution are listed in Table 6.6. According to this table, 166 EOLPs are disassembled in order to meet the component demands; 11 EOLPs are destructively disassembled to obtain parts to recycle, 200 EOLPs are remanufactured to satisfy the product demands, and the rest are left untouched.

Table 6.6 Disassembled and Remanufactured EOLPs

Activity	EOLP serial numbers
Disassembly	1, 2, 3, 5, 12, 16, 18, 20, 24, 26, 28, 29, 33, 36, 39, 43, 44, 45, 46, 50, 52, 53, 54, 60, 63, 65, 67, 68, 69, 72, 73, 74, 75, 76, 77, 81, 82, 84, 85, 86, 88, 89, 91, 96, 97, 102, 109, 111, 112, 114, 116, 117, 121, 125, 129, 131, 133, 134, 136, 137, 138, 140, 142, 143, 145, 149, 151, 152, 154, 155, 160, 164, 169, 170, 171, 174, 175, 178, 186, 188, 189, 191, 193, 196, 204, 207, 208, 209, 210, 215, 216, 217, 218, 221, 223, 224, 231, 234, 235, 245, 250, 251, 252, 254, 258, 261, 262, 264, 266, 267, 284, 287, 291, 294, 296, 297, 304, 305, 307, 308, 309, 310, 313, 316, 317, 320, 321, 325, 328, 329, 331, 332, 336, 337, 339, 342, 343, 349, 353, 355, 357, 359, 362, 363, 364, 365, 367, 369, 370, 371, 373, 374, 376, 378, 379, 380, 381, 383, 389, 390, 392, 393, 396, 397, 398, 400 (166 EOLPs)
Destructive disassembly	61, 107, 126, 139, 144, 177, 253, 263, 327, 335, 351 (11 EOLPs)
Remanufacturing	4, 6, 7, 8, 9, 10, 11, 13, 14, 15, 17, 19, 21, 22, 23, 25, 27, 30, 31, 32, 34, 35, 37, 38, 40, 41, 42, 47, 48, 49, 51, 55, 56, 57, 58, 59, 62, 64, 66, 70, 71, 78, 79, 80, 83, 87, 90, 92, 93, 94, 95, 98, 99, 100, 101, 103, 104, 105, 106, 108, 110, 113, 115, 118, 119, 120, 122, 123, 124, 127, 128, 130, 132, 141, 146, 147, 148, 150, 153, 156, 157, 158, 159, 161, 162, 163, 165, 166, 167, 168, 172, 173, 176, 179, 180, 181, 182, 183, 184, 185, 187, 190, 192, 194, 195, 197, 198, 199, 200, 201, 202, 203, 205, 206, 211, 212, 213, 214, 219, 222, 225, 226, 227, 228, 229, 230, 232, 233, 236, 237, 238, 239, 240, 241, 242, 243, 244, 246, 247, 248, 249, 255, 256, 257, 259, 260, 265, 268, 269, 270, 271, 272, 273, 274, 275, 276, 277, 278, 279, 280, 281, 282, 283, 285, 286, 288, 289, 290, 292, 293, 295, 298, 299, 300, 301, 302, 303, 306, 311, 312, 314, 315, 318, 319, 322, 323, 324, 326, 330, 333, 334, 338, 340, 341, 344, 345, 346, 347, 348, 350, 352, 354, 356, 358, 360, 361, 366, 372, 375, 377, 382, 384, 385, 386, 387, 388, 391, 394, 395, 399 (200 EOLPs)

Small samples of detailed disassembly (6 of 166 rows) and remanufacturing (6 of 200 rows) plans are given in Tables 6.7 and 6.8, respectively. For example, according to Table 6.7, when EOLP 1 is disassembled, components 1 and 9 are determined to be of low quality; components 4 and 8 are of medium quality, while components 3 and 10 are evaluated as high quality components. Similarly, according to Table 6.8, EOLP 4 is remanufactured to produce a medium quality product of type 1.

Optimal quantities of recycled nonreusable (nonfunctional and extremely low quality), low, medium, and high quality components are given in Table 6.9.

6.7 Conclusions

Awe inspiring growth in production has started to take its toll on the earth causing several global environmental problems. This has drawn a lot of public attention and has led to the birth of RL and ECMPRO that have recently become hot areas of

Table 6.7 Quality levels in which the disassembled components are placed

EOLP ID#	Components									
	1	2	3	4	5	6	7	8	9	10
1	L	–	H	M	–	–	–	M	L	H
2	L	L	–	H	L	H	–	M	–	H
3	M	M	L	H	–	–	–	–	–	–
.	.	.	.	.	.	.	.	.	.	.
.	.	.	.	.	.	.	.	.	.	.
.	.	.	.	.	.	.	.	.	.	.
397	H	H	–	–	–	–	–	–	M	H
398	L	L	M	–	–	–	–	H	H	M
400	M	M	H	L	–	–	M	H	–	L

L Low, *M* Medium, and *H* High

Table 6.8 Product types and quality levels of remanufactured EOLPs

EOLP ID#	Model	Quality level
4	1	Medium
6	3	High
7	8	High
.	.	.
.	.	.
.	.	.
394	7	Low
395	5	High
399	4	Low

Table 6.9 Optimal recycling quantities

Quality levels		Components									
		1	2	3	4	5	6	7	8	9	10
	Nonreusable	41	42	50	86	35	48	44	63	70	72
	Low	0	0	0	14	0	17	0	46	21	7
	Medium	0	0	0	0	0	0	0	0	0	0
	High	0	0	0	0	0	0	0	0	0	0

attention among both researchers and practitioners. However, uncertainty about the quality, quantity, and timing of the returned products complicates the planning of product recovery operations and makes it difficult to impose any quality standard.

Life cycle monitoring devices, such as sensors and RFID tags, bring clarity to RL operations and mitigate optimal recovery planning and quality assurance by providing much needed product usage information.

In this chapter, a RTO system for EOL SEPs was proposed. In the RTO system, collected life cycle information was used to determine the quality status of each and every component in inventory. Using this information, an IP model minimizing the total cost and satisfying the quality-based demands was constructed. Total cost was defined as the sum of total destructive and nondestructive disassembly, remanufacturing, recycling, and outside procurement costs. The model provided its users with the serial numbers of EOLPs to be subjected to aforementioned recovery processes and a detailed list of operations to be performed at the item level. A numerical case was considered to illustrate the model's application.

As a future research, computational complexity analysis of the model can be performed and various solution methodologies can be investigated.

References

1. Nikolaidis Y (2009) A modelling framework for the acquisition and remanufacturing of used products. Int J Sustain Eng 2(3):154–170
2. Chouinard M, D'Amours S, A-Kadi D (2008) A stochastic programming approach for designing supply loops. Int J Prod Econ 113(2):657–677
3. Nenes G, Nikolaidis Y (2012) A multi–period model for managing used products returns. Int J Prod Res 50(5):1360–1376
4. van Wassenhove LN, Zikopoulos C (2010) On the effect of quality overestimation in remanufacturing. Int J Prod Res 48(18):5263–5280
5. Dolgui A, Proth J-M (2008) RFID technology in supply chain management: state of the art and perspectives. In: Proceedings of the 17th international federation of automatic control world congress, Seoul, South Korea
6. Gungor A, Gupta SM (1999) Issues in environmentally conscious manufacturing and product recovery: a survey. Comput Ind Eng 36(4):811–853
7. Ilgin MA, Gupta SM (2010) Environmentally conscious manufacturing and product recovery (ECMPRO): a review of the state of the art. J Environ Manag 91(3):563–591
8. Kongar E, Gupta SM (2002) A multi-criteria decision making approach for disassembly-to-order systems. J Electron Manuf 11(2):171–183
9. Kongar E, Gupta SM (2009) Solving the disassembly-to-order problem using linear physical programming. Int J Math Oper Res 1(4):504–531
10. Kongar E, Gupta SM (2009) A multiple objective tabu search approach for end-of-life product disassembly. Int J Adv Oper Manag 1(2–3):177–202
11. Gupta SM, Imtanavanich P, Nakashima K (2010) Using neural networks to solve a disassembly-to-order problem. Int J Biomed Soft Comput Hum Sci (Special Issue on Total Operations Management) 15(1):67–71
12. Langella IM (2007) Heuristics for demand-driven disassembly planning. Comput Oper Res 34(2):552–577
13. Inderfurth K, Langella I (2006) Heuristics for solving disassemble-to-order problems with stochastic yields. OR Spectr 28(1):73–99
14. Kongar E, Gupta SM (2006) Disassembly to order system under uncertainty. Omega 34(6):550–561
15. Gupta SM, Taleb KN (1994) Scheduling disassembly. Int J Prod Res 32(8):1857–1866
16. Lambert AJD, Gupta SM (2005) Disassembly modelling for assembly maintenance, reuse, and recycling. CRC Press, Boca Raton
17. Taleb KN, Gupta SM (1997) Disassembly of multiple product structures. Comput Ind Eng 32(4):949–961
18. Taleb KN, Gupta SM, Brennan L (1997) Disassembly of complex products with parts and materials commonality. Prod Plan Control 8(3):255–269
19. Tang Y, Zhou M, Zussman E, Caudill R (2002) Disassembly modeling, planning, and application. J Manuf Syst 21(3):200–217

20. Veerakamolmal P, Gupta SM (1999) Analysis of design efficiency for the disassembly of modular electronic products. J Electron Manuf 9(1):79–95
21. Ilgin MA, Gupta SM (2012) Remanufacturing modeling and analysis. CRC Press, Boca Raton
22. Behret H, Korugan A (2009) Performance analysis of a hybrid system under quality impact of returns. Comput Ind Eng 56(2):507–520
23. Das K, Chowdhury AH (2012) Designing a reverse logistics network for optimal collection, recovery and quality-based product-mix planning. Int J Prod Econ 135(1):209–221
24. Denizel M, Ferguson M, Souza G (2010) Multiperiod remanufacturing planning with uncertain quality of inputs. IEEE Trans Eng Manag 57(3):394–404
25. Pochampally KK, Gupta SM (2006) Total quality management (TQM) in a reverse supply chain. In: Proceedings of the SPIE international conference on environmentally conscious manufacturing VI, Boston
26. Kim YS (2009) Quality embedded intelligent remanufacturing. Ph D thesis, LICP, EPFL, Lausanne
27. El Saadany AMA (2009) inventory management in reverse logistics with imperfect production, learning, lost sales, subassemblies, and price/quality considerations. Doctor of Philosophy Doctoral Dissertation, Department of Mechanical Engineering, Ryerson University, Toronto, Ontario
28. Chiu-Wei C-C (2010) Economics of cost of quality for green manufacturing life cycle assessment approach. Ph D thesis, Department of Industrial Engineering, Texas Tech University
29. Jun H-B, Kiritsis D, Xirouchakis P (2007) Research issues on closed-loop PLM. Comput Ind 58(8–9):855–868
30. Schmidt A, van Laerhoven K (2001) How to build smart appliances? IEEE Pers Commun 8(4):66–71
31. Soroor J, Tarokh MJ, Shemshadi A (2009) Initiating a state of the art system for real-time supply chain coordination. Eur J Oper Res 196(2):635–650
32. Cheng F-T, Huang G-W, Chen C-H, Hung M-H (2004) A generic embedded device for retrieving and transmitting information of various customized applications. In: Proceedings of the IEEE international conference on robotics and automation, New Orleans, LA
33. Chang C-Y, Hung S-S (2012) Implementing RFIC and sensor technology to measure temperature and humidity inside concrete structures. Constr Build Mater 26(1):628–637
34. Meyer GG, Främling K, Holmström J (2009) Intelligent products: a survey. Comput Ind 60(3):137–148
35. Karlsson B (1997) A distributed data processing system for industrial recycling. In: Proceedings of IEEE instrumentation and measurement technology conference (IMTC) 'sensing, processing, networking', Ottawa, Ontario, Canada
36. Petriu EM, Georganas ND, Petriu DC, Makrakis D, Groza VZ (2000) Sensor-based information appliances. IEEE Instrum Meas Mag 3(4):31–35
37. Scheidt L, Shuqiang Z (1994) An approach to achieve reusability of electronic modules. In Proceedings of IEEE international symposium on electronics and the environment, San Francisco, CA
38. Klausner M, Grimm WM, Hendrickson C, Horvath A (1998) sensor-based data recording of use conditions for product takeback. In: Proceedings of IEEE international symposium on electronics and the environment, Oak Brook, IL
39. Klausner M, Grimm WM, Horvath A (2000) Sensor-based data recording for recycling: a low-cost technology for embedded product self-identification and status reporting. In: Goldberg LH, Middleton W (eds) Green electronics/green bottom line. Butterworth-Heinemann, Woburn, pp 91–101
40. Yang X, Moore P, Chong SK (2009) Intelligent products: from lifecycle data acquisition to enabling product-related services. Comput Ind 60(3):184–194
41. Vadde S, Kamarthi SV, Gupta SM, Zeid I (2008) Product life cycle monitoring via embedded sensors. In: Gupta SM, Lambert AJD (eds) Environment conscious manufacturing. CRC Press, Boca Raton, pp 91–104

42. Ilgin MA, Gupta SM (2010) Comparison of economic benefits of sensor embedded products and conventional products in a multi-product disassembly line. Comput Ind Eng 59(4):748–763
43. Ilgin MA, Gupta SM (2011) Recovery of sensor embedded washing machines using a multi-Kanban controlled disassembly line. Robotics Comput Integr Manuf 27(2):318–334
44. Ilgin MA, Gupta SM (2010) Evaluating the impact of sensor-embedded products on the performance of an air conditioner disassembly line. Int J Adv Manuf Technol 53(9–12):1199–1216
45. Ilgin MA, Gupta SM (2011) Performance improvement potential of sensor embedded products in environmental supply chains. Resour Conserv Recycl 55(6):580–592
46. Luttropp C, Johansson J (2010) Improved recycling with life cycle information tagged to the product. J Clean Prod 18(4):346–354
47. Curran T, Williams ID (2012) A zero waste vision for industrial networks in Europe. J Hazard Mater 207–208:3–7
48. Parlikad AK, McFarlane D (2007) RFID-based product information in end-of-life decision making. Control Eng Pract 15(11):1348–1363
49. Gonnuru VK (2010) Radio-frequency identification (RFID) integrated fuzzy based disassembly planning and sequencing for end-of-life products. Masters Thesis, Mechanical Engineering, The University of Texas at San Antonio, San Antonio
50. Kulkarni AG, Parlikad AKN, McFarlane DC, Harrison M (2005) Networked RFID systems in product recovery management. In: Proceedings of IEEE international symposium on electronics and the environment, New Orleans, LA
51. Zhou S, Ling W, Peng Z (2007) An RFID-based remote monitoring system for enterprise internal production management. Int J Adv Manuf Technol 33(7–8):837–844
52. Engel SJ, Gilmartin BJ, Bongort K, Hess A (2000) prognostics, the real issues involved with predicting life remaining. In: Proceedings of IEEE Aerospace Conference, Big Sky, MT
53. Lee BS, Chung HS, Kim K-T, Ford FP, Andersen PL (1999) Remaining life prediction methods using operating data and knowledge on mechanisms. Nucl Eng Des 191(2):157–165
54. Middendorf A, Griese H, Grimm WM, Reichl H (2003) Embedded life-cycle information module for monitoring and identification of product use conditions. In: Proceedings of 3rd international symposium on environmentally conscious design and inverse manufacturing (EcoDesign), Tokyo, Japan
55. Middendorf A, Reichl H, Griese H (2005) Lifetime estimation for wire bond interconnections using life-cycle-information modules with implemented models. In: Proceedings 4th international symposium on environmentally conscious design and inverse manufacturing (EcoDesign), Tokyo, Japan
56. Rugrungruang F (2008) An integrated methodology for assessing physical & technological life of products for reuse. Ph D thesis, Life Cycle Engineering & Management Research Group, School of Mechanical and Manufacturing Engineering, The University of New South Wales, New South Wales, Australia
57. Wang W, Zhang W (2008) An asset residual life prediction model based on expert judgments. Eur J Oper Res 188(2):496–505
58. Herzog MA, Marwala T, Heyns PS (2009) Machine and component residual life estimation through the application of neural networks. Reliab Eng Syst Saf 94(2):479–489
59. Mazhar MI, Kara S, Kaebernick H (2007) Remaining life estimation of used components in consumer products: life cycle data analysis by Weibull and artificial neural networks. J Oper Manag 25(6):1184–1193
60. Byington CS, Watson M, Edwards D (2004) Data-driven neural network methodology to remaining life predictions for aircraft actuator components. In: Proceedings of the IEEE aerospace conference

Chapter 7
An RFID Integrated Quality Management System for Reverse Logistics Networks

Anjali Awasthi and S. S. Chauhan

Abstract Quality management is vital for reverse logistics networks to improve productivity and reduce waste. Every year, several tons of products are wasted due to lack of data collection, poor information management, shortage of technological solutions, and lack of quality management knowledge in this area [1–3]. In this chapter, we present an RFID integrated quality management system for reverse logistics networks. The proposed tool integrates four functionalities namely data collection, analytical processing, quality monitoring, and recommendations generation. An RFID integrated network assures accurate data collection in real time. The data are then analytically processed and subject to different quality management techniques for quality monitoring and analysis. A numerical case study is provided to demonstrate the application of the proposed quality management system for reverse logistics networks.

7.1 Introduction

The importance of reverse logistics (RL) has been growing over the years and is expected to become even more important in the coming years due to its contribution in greening of supply chains, wise usage of natural resources, and protection of environment. RL involves collection of returned and discarded goods from

A. Awasthi (✉)
CIISE – EV 7.640, Concordia University, Montreal,
QC H3G2W1, Canada
e-mail: awasthi@ciise.concordia.ca

S. S. Chauhan
Decision Sciences, Molson School of Business, Concordia University, Montreal,
QC H3G2W1, Canada
e-mail: sschauha@alcor.concordia.ca

Y. Nikolaidis (ed.), *Quality Management in Reverse Logistics*,
DOI: 10.1007/978-1-4471-4537-0_7, © Springer-Verlag London 2013

collection points (houses, enterprises, stores, collection centers, etc.) to the treatment facilities (or processing centers), where they are sorted and separated into different classes of products depending upon the next stage of treatment (disposal, recycling or reprocessing, and redistribution to a secondary market). Rogers and Tibben-Lembke [4] define RL as

> the process of planning, implementing, and controlling the efficient, cost-effective flow of raw materials, in-process inventory, finished goods, and related information from the point of consumption to the point of origin for the purpose of recapturing value or proper disposal.

van Hillegersberg et al. (2001) distinguish RL operations from traditional logistics operations in terms of consumers' behavior which introduces uncertainties in the quality, quantity, and timing of product returns. Jun and Kim [5] present the following main characteristics of RL:

1. There is an uncertainty of returned time of a product.
2. There is an uncertainty of quality of a returned product, i.e., uncertainty of recovered value.
3. There is an uncertainty of configurations of parts or components of returned products.
4. There is an uncertainty of locations.

Considering the above uncertainties, it is very important to capture accurate and right amount of information to enable timely product recovery in RL networks. Ferguson and Browne [6] classify the information required for efficient product recovery into six categories. These are (a) product related (b) location related (c) utilization related (d) legislative information (e) market information, and (f) process information. Parlikad et al. [7] classify the necessary information for product disassembly into two groups: internal and external. The internal information consists of design information, reliability information, disassembly information, production information, location information, and life cycle information. The external information consists of market information, legislative information, process-related information, and corporate policies.

Most of the RL operations (collection, sorting, assembly, etc.) involve manual recording of data which contains human errors or inconsistency often caused due to duplication of tasks, lack of technological equipment, and IT tools. The update of data is also not simultaneous with the arrival times of goods at the points of collection which can significantly affect the quantity and quality of product retrieval at later stages. Using the quality cost model, it would be imperative to invest in the prevention costs at early stages to avoid appraisal, internal, and external failure costs of products down the chain at later stages. In order to overcome this issue, several technological solutions such as Internet, wireless mobile telecommunication technologies, product identification technologies such as auto-id, GPS etc., are being investigated. RFID (Radio Frequency Identification) is one of such solutions that can be used for item tracing and real-time data management in RL networks to ensure product quality for better recovery at later stages.

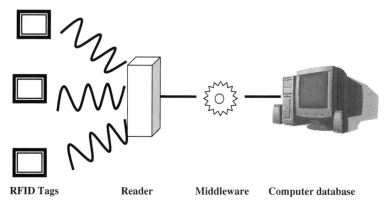

RFID Tags   Reader   Middleware   Computer database

Fig. 7.1 Anatomy of an RFID system

The rest of the sections are organized as follows. In Sect. 7.2, we present the literature review. Section 7.3 provides the details of the proposed RFID integrated quality management system. Section 7.4 presents a numerical case study of the proposed quality management system. Finally, we present the conclusions in Sect. 7.5.

7.2 Literature Review

7.2.1 What is RFID?

RFID is a standard term to describe technologies that utilize radio waves to capture and identify data. RFID uses wireless technology to convey data between the microchip-embedded tags and readers. The tags, consisting of a microchip and an antenna, are attached to the objects that need identification. The reader, using one or more antennae, reads the data held on the microchip. The data are then communicated to a computer database or information management system using a middleware [8, 9]. Figure 7.1 presents the anatomy of an RFID system.

The RFID tags can be classified into active, passive, and semi-passive [8, 9]. The active RFID tags have a power source such as an onboard battery attached to the tag. The typical read range of an active tag is usually between 20 and 100 m and the batteries last up to several years. The passive RFID tags do not have a power source, but instead generate power through induction. The antennae on both the tag and the reader usually have a small coil, and when placed near one another, form an electromagnetic field which provides power to the tag. Thus, if a reader is not present, the passive tag cannot communicate any data. Active tags can communicate in the absence of a reader. The typical range of a passive tag ranges from a few inches to 10 m. The semi-passive tags have their own battery which is used to run the microchip. They may only respond to incoming transmissions and communicate using the power provided on the reader. The maximum range of a semi-passive tag is 100 m.

7.2.2 RFID and RL

RFID has been applied in various sectors including manufacturing, retail, military, tollgates, airports, construction etc., for product identification, inventory management, security checks, and tracking of items [10, 11], Lu and Huang [12], [13–16]. Tajima [17] demonstrates the strategic value of RFID in supply chain management. Saygin et al. [18] present a systems approach to viable RFID implementation in the Supply Chain.

Application of RFID in RL is an important topic; however, very few studies have been conducted in this area. Lee and Chan [19] develop an RFID-based RL system. Jung et al. [20] integrate RFID with quality assurance system. Langer et al. [21] assess the impact of RFID on return center logistics. Lu et al. [22] study RFID-based information management in the automotive plastic recycling industry. Jun and Kim [5] present research issues for enhancing the productivity of RL using information technologies. Jayaraman et al. [1] study the role of information technology and collaboration in RL supply chains. Dekker et al. [3] and Flapper et al. [23] present a compilation of practical examples and case studies in the area of RL. Trappey et al. [24] use genetic algorithm for dynamic performance evaluation for RFID-based RL management.

7.2.3 Challenges in RFID Implementation

Using RFID technologies in the RL area comes with many benefits such as information visibility in time, location, and product status. In particular, it can increase the number of recovered products by simplifying the operations of collecting and storing, disassembling, and facilitating efficient product recovery decision [2, 25]. However, challenges that need to be addressed in RFID implementation are presented as follows:

- Number of tags and readers to be used
- Number of read points, and their locations
- Interface of RFID system with existing quality control software
- Level of worker training required
- Security issues in sharing RFID information between companies
- Standardization related to adoption of RFID technologies by participating companies involved in RL
- Interoperability issues in sharing product information between companies
- Cost of RFID implementation

Some of these challenges have been addressed in [8–9], [12–13], [15–17].

7.2.4 Motivation for Applying RFID for Quality Management in RL

This chapter is motivated by the application of RFID technology in manufacturing environments to avoid losses occurring from high return rate of poor quality products. Using the RFID technology, it is possible:

- to have accurate inventory of products;
- to know the bill of materials (BOM) of products, thereby enabling easier identification of defective parts or subcomponents;
- to trace locations from and to which the returned goods are shipped, know the reason and time of return;
- to update product information on the chip, etc., which is not possible with technologies like barcode;
- to maintain a centralized database where the information collected from RFID tags of returned products can be stored and shared across multiple locations in real time for quality control purposes;
- to perform advance recall of products that have used common defective components identified using BOM and avoid extra costs involved in transportation handling of returned products returned at later stages.

7.3 RFID Integrated Quality Management System

The proposed RFID integrated quality management system comprises four modules. These modules are:

- Data collection module
- Analytical processing module
- Quality monitoring module
- Recommendations generation module

Figure 7.2 presents the various components of the decision-making tool.

7.3.1 Data Collection Module

The first module called the data collection module employs an RFID integrated network to collect data. The RFID tags can be implemented at item level, case level, or pallet level. The item tag usually contains the product id and the case/pallet tag contains information on the number and types of items contained in it. The data collection in RL networks happens at three stages, namely collection, transportation, and disposition. These stages are explained as follows:

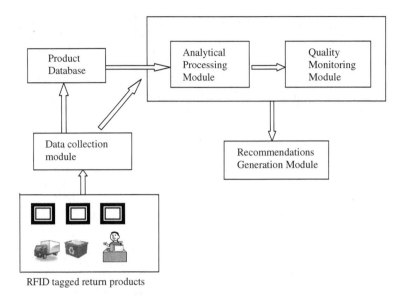

RFID tagged return products

Fig. 7.2 Components of the decision support system

7.3.1.1 Collection Stage

This stage involves retrieval of goods from customers. The returns include end-of-use and end-of-life items. The retrieval can happen at stores, collection centers, or consumer locations. The store level returns comprise product recalls, inventory returns, warranty returns, core returns, reusable containers, damaged goods, seasonal items, hazardous materials, and stock adjustments. The collection center retrievals comprise product recalls, warranty returns, inventory returns, core returns, reusable container returns, damaged goods, seasonal items, and hazardous materials. The consumer returns are collected from the ultimate customer and comprise recyclable goods, damaged goods, and product recalls.

7.3.1.2 Transportation Stage

This stage involves transportation of goods from consumer locations, collection centers, and stores to treatment facilities. The trucks containing tagged pallets, cases, or items pass through a reader and the stored item information is communicated to product database through a middleware for updating returned/recalled goods inventory.

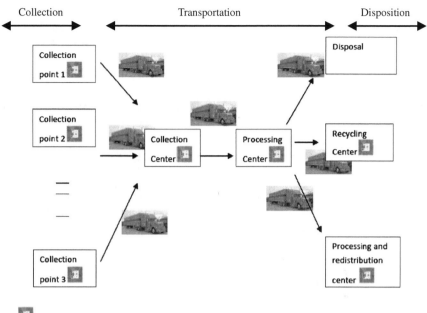

Fig. 7.3   RFID enabled network for data collection in RL

7.3.1.3 Disposition Stage

The tagged items upon arrival at the treatment facilities or processing centers are sorted into different classes depending upon their quality and recoverability. The sorted products are then visually inspected to identify the defect types and directed to different channels of repair, reassembly, remanufacturing, redistribution, or disposal.

Figure 7.3 presents a graphical illustration of the deployment of RFID tags at the three stages in a RL network.

7.3.2 Analytical Processing Module

The role of the analytical processing module is to read the information from the product database, analyze the product quality, and determine the next steps for processing. Depending upon the reception point of returned goods, the analytical processing can be done at (a) customer locations (b) collection centers, or (c) departmental/retail stores.

For the returned goods received at the customer locations, the analytical processing module uses return product data to determine which customer locations provide high volumes of returned products, the types of return products, period of return, the condition of returned products, and so on. Data collection through RFID integrated network ensures accurate availability of information on these parameters, which can be used for location planning of collection points, number of collection points to be installed, etc., resulting in better management of customer demand and yield of returned products.

For the returned goods received at collection centers, the analytical processing module keeps track of information on the yield rate of different collection centers, types of return products, etc., which can be used for capacity management of collection centers, allocation of vehicle resources for returned product collection and dispatch to recycling facilities, scheduling of driver resources, and so on.

For the returned goods received at departmental/retail stores treatment facility channel, accurate monitoring and record keeping of return products data result in better inventory management, improved customer satisfaction, increased awareness about recyclability of returned products, and so on. The stores can also perform informed supplier selection based on history data of returned products to retain customer loyalty.

7.3.3 Quality Monitoring Module

The role of quality monitoring module is to perform statistical analysis of collected data to analyze product, process, or service quality. Monitoring can be done at customer locations, collection centers, stores to study variability in quality and quantity of return products at different life cycle stages over different periods of time. Graphical tools, such as histograms, run charts, and descriptive statistics, are used to display product quality. For process quality, statistical control charts are used. Different types of attribute or variable control charts can be deployed based on the type, quality, and quantity of information available about the products.

7.3.4 Recommendations Generation Module

This module provides recommendations for improving product, process, or service quality based on the results provided by the analytical processing and quality monitoring modules. For out-of-control situations, an out-of-control action plan [26] is used to generate recommendations for improvement. In situations where control charts are not used, other techniques such as cause-and-effect diagram, root cause analysis, 5 whys technique, tree diagram, etc., are used to generate solutions for the quality problem under investigation.

It can be seen from the functionalities of the above four modules that the proposed RFID integrated quality management system is capable of recording product data in a timely, correct, and efficient manner for further processing through the other three modules to assure product, process, and service quality to customers. This can help save treatments costs, reduce scrap, and improve product restorability in RL networks.

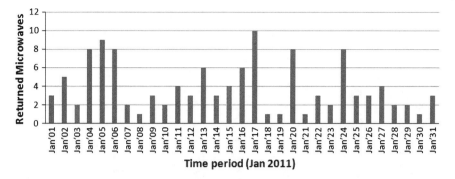

Fig. 7.4 Distribution of returned microwave ovens from Jan 1 to 31, 2011

7.4 Numerical Application

In this section, we present the application of the proposed RFID integrated quality management system for ABC Inc., which has 10 retail stores across different locations in Canada. ABC sells different types of electronic appliances for household cooking such as microwave ovens, juicers, toasters, and so on. The name of the company and the original data have been modified for confidentiality reasons.

7.4.1 Data Collection

The material flow of returned items at ABC can be categorized into "inbound" and "outbound". The "inbound" flow commences from the reception of goods at the reception desk until storage in the inventory. The "outbound" flow commences when the inventoried return items are sent for (a) recovery (b) recycling, or (c) disposition. The recovery happens when the returned items are sent back to the manufacturer/supplier, or sold to individuals in small quantities at lower prices. In recycling, the returned products are sold to recycling organizations for further processing such as remanufacture, reassembly, and so on. In disposition, the returned products with zero resale value are scrapped. Customers dissatisfied with product quality return the products to retail stores. Using the embedded RFID tag, the store retrieves the identification details of the returned products in the database. If the product was under warranty, or is defective, or the customer is not satisfied with the product for any other reason, it is returned and the customer is refunded.

Figure 7.4 shows the graphical distribution of the total number of 121 returned microwaves for Jan 01–31, 2011 at 10 retail locations of ABC. It can be seen that the highest number of returns (10) took place on Jan 17 followed by Jan 05, 04, 06, 20, and 24. The lowest number of returns, namely one, took place during Jan 08, 18, 19, and 30.

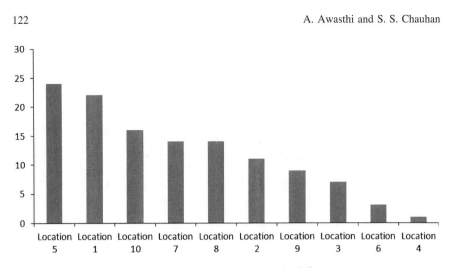

Fig. 7.5 Distribution of returned microwave ovens at the 10 ABC stores

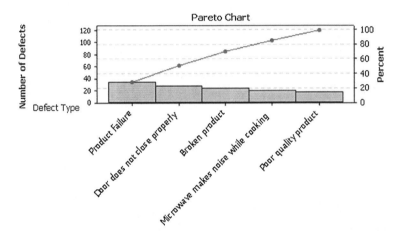

Fig. 7.6 Commonly reported reasons for microwave returns at ABC stores

The classification of the 121 returned items of January 2011 according to the store locations is shown in Fig. 7.5. Note that with the help of RFID, this information can be shared and monitored in real time across the 10 locations of ABC which is not possible with barcodes. It can be seen that location 5 has the highest number of returned microwaves, namely 24, followed by locations 1, 10, 7, and 8 respectively. Location 4 has the least number of microwave returns.

The product quality is monitored on different days at ABC Inc., to identify on which days greater number of product returns occurred so that the source problems can be identified. These include the supplier/manufacturer providing the poor quality product for those days or mishandling occurred at ABC itself. If the source of the problem is the supplier/manufacturer, then the defective products are returned to them under contract/guarantee conditions. The manufacturer/supplier on receipt of the defective product can then perform a detailed analysis of causes such as transportation, warehousing, or manufacturing, etc., and take remedial action. If the source of the problem is ABC itself, then appropriate actions for quality control are taken at its end. All these decisions can be made in a proactive manner with the help of RFID.

Figure 7.6 presents the Pareto chart showing the distribution of defects or nonconformities in the returned products according to their types. It can be seen in Fig. 7.6 that "product failure" is the most commonly reported defect followed by "door does not close properly", "broken product", "microwave makes noise while cooking", and "poor quality". Since "product failure" is most likely to occur due to manufacturing/production error, ABC returns all the defective products of this category to the manufacturer (if they are directly procured by them) or to the supplier to investigate the possible reasons and get refunds.

7.4.2 Analytical Processing

Table 7.1 presents the returned microwave oven data across the 10 store locations of ABC. Please note that the data provided in Table 7.1 are the detailed representation of returned items presented in Fig. 7.4 for the month of January, 2011. It was randomly chosen from the original data set (confidential) of ABC Inc., for illustration purposes.

It can be seen that the rate of returns is not constant but varies over different days and also across different locations. Therefore, we have variable return product data for the microwave ovens. Assuming a constant sample size of 100 from which the returned products are assessed for each of the 10 locations of ABC Inc., per day, the control chart that can be used for monitoring quality at ABC Inc., is the c-control chart. The control limits for the c-chart for the 10 store locations are presented in the last three rows of Table 7.1.

7.4.3 Quality Monitoring

Figure 7.7 presents the graphical display of the results of the c-chart for the month of January 2011 across the 10 store locations of ABC. It can be seen that the highest number of out-of-control points, namely four, is observed for locations 5, 8, and 1, followed by three out-of-control points for location 9, two out-of-control points for locations 2, 6, 7, and 10. The least number of out-of-control points, i.e., one, are observed for locations 3 and 4. Therefore, most attention is required for quality monitoring and control at locations 5, 8, and 1. For each of the out-of-control points at its 10 stores ABC, a more detailed investigation should be done to identify the sources of the problem and take appropriate action.

Figure 7.8 presents the results of the overall conformities observed per day irrespective of the store locations of ABC. The LCL = 0, CL = 3.9, and UCL = 9.83. It can be seen that there is one out-of-control point for Jan 17. Other days on which high number of product returns are observed are Jan 04, 06, 20, and

Table 7.1 Upper and lower control limits for the returned microwave oven data

Date	Locations										Returned microwaves (c)
	L1	L2	L3	L4	L5	L6	L7	L8	L9	L10	
Jan'01	0	0	0	0	0	0	0	0	3	0	3
Jan'02	0	0	5	0	0	0	0	0	0	0	5
Jan'03	0	0	0	0	2	0	0	0	0	0	2
Jan'04	0	0	0	0	8	0	0	0	0	0	8
Jan'05	0	0	0	0	0	0	9	0	0	0	9
Jan'06	0	8	0	0	0	0	0	0	0	0	8
Jan'07	0	0	0	0	0	2	0	0	0	0	2
Jan'08	0	0	1	0	0	0	0	1	0	0	1
Jan'09	0	0	0	0	0	0	0	3	0	0	3
Jan'10	0	0	0	0	0	0	0	0	0	0	2
Jan'11	0	0	0	0	4	0	0	0	0	0	4
Jan'12	0	0	0	0	0	0	3	0	0	0	3
Jan'13	0	0	0	0	6	0	0	0	0	0	6
Jan'14	0	0	0	0	0	0	0	3	0	0	3
Jan'15	0	0	0	0	4	0	0	0	0	0	4
Jan'16	6	0	0	0	0	0	0	0	0	0	6
Jan'17	10	0	0	0	0	0	0	0	0	0	10
Jan'18	0	0	0	1	0	0	0	0	0	0	1
Jan'19	0	0	0	0	0	1	0	0	0	0	1
Jan'20	0	0	0	0	0	0	0	0	0	8	8
Jan'21	0	1	0	0	0	0	0	0	0	0	1
Jan'22	0	0	0	0	0	0	0	3	0	0	3
Jan'23	0	2	0	0	0	0	0	0	0	0	2
Jan'24	0	0	0	0	0	0	0	0	0	8	8
Jan'25	3	0	0	0	0	0	0	0	0	0	3
Jan'26	3	0	0	0	0	0	0	0	0	0	3
Jan'27	0	0	0	0	0	0	0	0	4	0	4
Jan'28	0	0	0	0	0	0	0	0	2	0	2
Jan'29	0	0	0	0	0	0	2	0	0	0	2
Jan'30	0	0	1	0	0	0	0	0	0	0	1
Jan'31	0	0	0	0	0	0	0	3	0	0	3
UCL	3.23	2.14	1.65	0.57	3.41	1.03	2.46	2.36	1.91	2.67	9.83
CL	0.709	0.35	0.23	0.032	0.77	0.096	0.45	0.419	0.29	0.516	3.903
LCL	0	0	0	0	0	0	0	0	0	0	0

24. ABC Inc., can look into the inventory received on these days in more detail to identify the sources of poor quality and take appropriate action.

7.4.4 Recommendations Generation

Using the return product data and the reasons for the return of products, it was found that the most common reasons for the return of products were "product failure" and "door does not close properly". Since, these defects are likely to

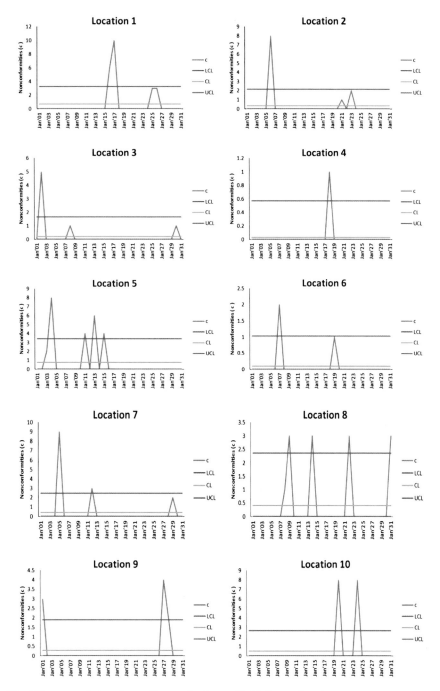

Fig. 7.7 c-control chart for quality monitoring at 10 ABC store locations

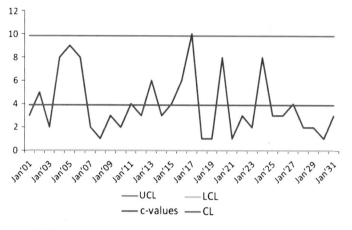

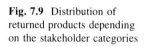

Fig. 7.8 *c*-control chart

Fig. 7.9 Distribution of returned products depending on the stakeholder categories

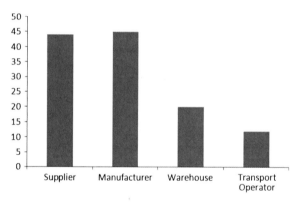

occur at the manufacturing stage or reception of poor quality product from the supplier; therefore, manufacturer and suppliers were considered to be most likely responsible for high return rate of microwaves on Jan 4, 5, 6, 17, 20, and 24, respectively. The other stakeholder categories are warehouse manager and transport operator (Fig. 7.9). Cause-and-effect diagram was used to identify the stakeholders responsible for product returns.

Therefore, based on the results from the Quality Management System, ABC Inc., can notify the manufacturers and suppliers about the poor quality of their products and ask for tighter inspection and quality control procedures before sending products. Upon receiving the retuned products from ABC, the product manufacturers/suppliers can do a thorough investigation of high return rate of its products due to product failure/poor quality. Due to the presence of embedded RFID chip in the return products, the manufacturer/supplier can look into the BOM and identify which component(s) is the root source of the problem. Then, the departments such as Inspection, Warehousing, Manufacturing, Transportation, etc., through which this component went through until final delivery to customer, are investigated. Once the erroneous component has been found, techniques like cause-and-effect diagram are used to identify the underlying reasons for their failure such as man, machine, methods, material, measurement, and so on. Based upon these reasons, the suppliers/manufacturers may do closed monitoring of quality control activities in their department to identify reasons for variations in product failure over time for different product types in different customer regions, etc., and take appropriate action to remedy the situation. The presence of accurate and timely information from RFID thereby reduces damage costs of goods by better tracking, reduction in false claims on quantities of goods dispatched/ received, and increased product recovery due to accurate demand and inventory management.

7.5 Conclusions and Future Work

In this chapter, we present an RFID integrated quality management system for RL networks. Quality assurance in these networks is very important to improve product restorability, reduce wastes, and conserve natural resources for a greener society and cleaner environment.

The proposed quality management system integrates four functionalities namely data collection, analytical processing, quality monitoring, and recommendations generation. An RFID integrated network is proposed to ensure accurate data tracking and collection in real time. The data are then analytically processed and subject to different quality management techniques for quality monitoring and analysis. A numerical case study is provided to demonstrate the application of the proposed product quality monitoring tool on RL networks.

The strength of the proposed tool can be attributed to the increased data accuracy, the online data collection, and the ability to generate recommendations for quality management in RL networks through the use of appropriate quality control techniques in real time. Due to the presence of RFID, it is possible to trace defective subcomponents in a product, identify other products in which this defective component was used, and therefore the manufacturer/supplier can do advance recall of its products and avoid extra costs involved in transportation, handling of returned products at later stages, thereby ensuring return of products to the right source locations and retaining loyalty and trust of the customer in the company.

Our future work involves development and application of the proposed tool for quality management decision-making in real-life instances in different types of RL networks such as electronics, aviation, and so on.

References

1. Jayaraman V, Ross AD, Agarwal A (2008) Role of information technology and collaboration in reverse logistics supply chains. Int J Logist Res Appl 1469-848X, 11(6):409–425
2. Payaro A (2004) The role of ICT in reverse logistics: a hypothesis of RFID implementation to manage the recovery process. In: Proceedings of the 2004 eChallenges Conference
3. Dekker R, Fleischmann M, Inderfurth K, Wassenhove LN, van (eds) (2004) Reverse logistics, quantitative models for closed-loop supply chains, VIII, 436, ISBN: 978-3-540-40696-9
4. Rogers DL, Tibben-Lembke RS (1999) Going backwards: reverse logistics trends and practices. Reverse Logistics Executive Council, Pittsburgh
5. Jun H-B., Kim J-G (2006) State-of-the art: research issues and framework for enhancing the productivity of reverse logistics using emerging information technologies http://appc.snu.ac.kr/files/1189_paper.pdf
6. Ferguson N, Browne J (2001) Issues in end-of-life product recovery and reverse logistics. Prod Plan Control 12(5):534–547
7. Parlikad AK, McFarlane D, Fleisch E, Gross S (2003) The role of product identity in end-of-life decision making, White paper, Auto-ID center, Cambridge
8. Kou D, Zhao K, Tao Y, Kou W (2006) RFID technologies andapplications. In: Kou W, Yesha Y (eds) Enabling technologies forwireless e-business. Springer, The Netherlands, pp 89–108
9. Beckner M, Simms M, Venkatesh R (2009) RFID background primer. Pro RFID in BizTalk Server 2009, Apress publications, 1–17
10. Abad E, Palacio F, Nuin M, González de Zárate A, Juarros A, Gómez JM, Marco S (2009) RFID smart tag for traceability and cold chain monitoring of foods: demonstration in an intercontinental fresh fish logistic chain. J Food Eng 93(4):394–399
11. Oztekin A, Pajouh FM, Delen D, Swim LK (2010) An RFID network design methodology for asset tracking in healthcare. Decis Support Syst 49(1):100–109
12. Lu W, Huang GQ, Li H (2010) Scenarios for applying RFID technology in construction project management. Autom Constr, Corrected Proof, Available online 28 October 2010
13. Müller-Seitz G, Dautzenberg K, Creusen U, Stromerede C (2009) Customer acceptance of RFID technology: Evidence from the German electronic retail sector. J Retail Consumer Serv 16(1):31–39
14. Ngai EWT, Suk FFC, Lo SYY (2008) Development of an RFID-based sushi management system: the case of a conveyor-belt sushi restaurant. Int J Prod Econ 112(2):630–645
15. Gaukler G, Seifert RW (2007) Applications of RFID in supply chain management. in: Jung H, Chen FF, Jeong B (eds) Trends in supply chain design and management: technologies and methodologies, Special Series in Advanced Manufacturing, Springer, London pp. 29–48
16. Chowdhury B, Chowdhury MU, D'Souza C (2008) Challenges relating to RFID implementation within the electronic supply chain management–a practical approach, software engineering, artificial intelligence, networking and parallel/distributed computing. Stud Comput Intell 149(2008):49–59
17. Tajima M (2007) Strategic value of RFID in supply chain management. J Purch Suppl Manag 13(4):261–273
18. Saygin C, Sarangapani J, Grasman S (2007) A systems approach to viable RFID implementation in the supply chain, trends in supply chain design and management. Springer Series in Advanced Manufacturing, Part I., 3–27
19. Lee CKM, Chan TM (2009) Development of RFID-based reverse logistics system. Expert Syst Appl 36(5):9299–9307
20. Jr Jung Lyu, Chang S-Y, Chen T-L (2009) Integrating RFID with quality assurance system – Framework and applications. Expert Syst Appl 36(8):10877–10882
21. Langer N, Forman C, Kekre S, Scheller-Wolf A (2007) Assessing the impact of RFID on return center logistics. Interfaces 37(6):501–514

22. Lu Z, Cao H, Folan P, Potter D, Browne J (2007) RFID-based information management in the automotive plastic recycling industry, information technologies in environmental engineering. Environ Sci Eng 8:397–408
23. Flapper Simme DP, Nunen Jo, A.E.E. van, van Wassenhove LN (eds) (2005) Managing closed-loop supply chains, XII, 214, ISBN: 978-3-540-40698-3***
24. Trappey Amy JC, Trappey Charles V, Chang-Ru Wu (2010) Genetic algorithm dynamic performance evaluation for RFID reverse logistic management. Expert Syst Appl 37(11):7329–7335
25. Parlikad AK, McFarlane DC, Kulkarni AG (2006) Improving product recovery decisions through product information. Innov Life Cycle Eng Sustain Dev 2:153–172
26. Montgomery DC (2009) Introduction to statistical quality control. Wiley Publications, ISBN 978-0-470-16992-6

Chapter 8
Cases of Damage in Third-Party Logistics Businesses

Berrin Denizhan and K. Alper Konuk

Abstract Although the goal of Third-Party Logistics (3PL) providers is to transport goods without incidents, cases of damage do occur making the control and prevention of damage one of the significant quality concerns of Closed-Loop Supply Chains (CLSC) and Reverse Logistics (RL) systems. The costs incurred through damage are often viewed as a necessary type of operating cost but there is no standard procedure for documenting and evaluating cases of damage. This chapter describes common types of damage sustained during the 3PL process and proposes a damage classification system which will improve the quality of 3PL services through an efficient tracking and evaluating system. This system will enable 3PL providers to identify and correct the systemic sources of incidents of damage. It also suggests that damages can be classified according to the location of the occurrence of damage, the type of damaged product, and the means by which damage is defined. Using a case study approach, this research analyzes incidents of damage that occurred at Borusan Logistics in Turkey, between 2005 and 2007. Based on this analysis, this study suggests that by designating its own risk expert to evaluate and manage cases of damage, a 3PL provider will benefit in several ways, as will the manufacturer of transported goods and the insurance company. These benefits include a more accurate and efficient damage evaluation process, expedited processing of insurance claims, and refinements to the stipulations of liability contracts. The proposed model may reduce cases of damage both by making patterns of damage visible and by clarifying appropriate corrective actions.

B. Denizhan (✉)
Department of Industrial Engineering, Sakarya University,
54178 Sakarya, Turkey
e-mail: denizhan@sakarya.edu.tr

K. Alper Konuk
Borusan Logistics, Gemlik, 16601 Bursa, Turkey
e-mail: akonuk@borusan.com

Y. Nikolaidis (ed.), *Quality Management in Reverse Logistics*,
DOI: 10.1007/978-1-4471-4537-0_8, © Springer-Verlag London 2013

8.1 Introduction

Reverse logistics (RL) includes the processing of merchandise that has been returned as a result of damage, seasonal inventory, restock, salvage, recalls, and excess inventory [1]. The main operational factors of RL systems are cost benefit analysis, transportation, warehousing, supply chain management, remanufacturing, recycling, and packaging. Each company should coordinate its RL activities with loads, networks, and inbound and outbound transportation services to take full advantage of those activities [2]. All these RL activities may contain a greater number of distribution flows than those in forward logistics. Consequently, this complexity of transportation needs results in the emergence of Third-Party Logistics (3PL) providers as significant players in the RL field, both for the reduction of complexity and the increase of economic advantage [3, 4].

Products are returned or discarded either because they do not function properly or because they have become redundant. Besides in-transit damage, other reasons for entry into the RL process include defective or faulty manufacturing and customer dissatisfaction [4, 5]. The initiation of a return can be arranged according the typical supply chain hierarchy, starting with manufacturing, proceeding from manufacturers to wholesaler/retailers and finally to customers/consumers, who in principle are going to use the products. Manufacturing returns include raw material surplus, quality-control returns, and production leftovers. Distribution returns include product recalls, commercial returns (e.g., unsold products, wrong/damaged deliveries), stock adjustments, and functional returns. Finally, customer returns include reimbursement guarantees, warranty returns, service returns (repairs and spare parts), end-of-use, and end-of-life returns [5].

A closer investigation of the nature of product damage and the operations in which damage takes place will enable 3PLs to improve service quality, reduce costs, and increase efficiency. This study uses an analysis of the instances of damage occurred in Borusan Logistics, a large 3PL and port service provider in Turkey, and examines the most frequent types of damage at 3PL providers, their causes, and possible methods for their prevention.

Borusan Logistics focuses on 3PL services and port management. Following the growth of operations both in volume and in geographic scope, Borusan Logistics extends logistics' services to major regional customers in specific sectors. A series of subsidiaries includes Borusan Logistics International Gulf FZE in the United Arab Emirates, Borusan Logistics International Algeria SPA, Borusan Logistics International Netherlands and Borusan Logistics International USA. Borusan Logistics International carries out Third-Party Integrated Logistics Services as a growing regional power in Ukraine, Rumania, Russia, Hungary, CIS countries, North Africa, Syria, Iraq and Lebanon, as well as, the Benelux countries.

The literature to date does not contain a comprehensive discussion of the theory of 3PL's damages. Currently, the costs incurred through damage are included in operating costs, and therefore are regarded as minor expenses that are uninsurable and unavoidable (at least in Borusan Logistics). However, condensing the diverse kinds of damages and the respective costs result in irregular and inaccurate recording of cases of damage.

This study investigates the real causes of damage in a 3PL and suggests a framework within which damages can be more accurately classified. In the next section, basic notions of 3PLs and RL are presented. In Sect. 8.3, notions of the damages and RL are introduced. Finally, recording and controlling the cases of damage with basic spreadsheet reports, as well as their proper organizational structure are suggested in Sect. 8.4.

8.2 Third-Party Logistics and Reverse Logistics

A 3PL provider is an external contractor who manages, controls, and delivers logistics activities on behalf of a shipper [6]. Although firms have relied for years upon 3PL providers to ship their products forward, they have only recently begun to rely on them for their RL activities [7]. RL entails the planning and the necessary processes for the transportation of goods from their final destination back to their point of origin, either for recapturing the remaining value or for disposal. Consequently, efficient, cost-effective, and specialized RL processes put value back into goods that might otherwise become worthless. More specifically, RL processes include the collection and return of industrial and technical products and parts; the recycling of packaging and pallets; the validation and control of green products; and the collection, processing, sorting, and marketing of waste and hazardous materials.

According to the 14th annual study on the State of Logistics Outsourcing (2009), many of the logistics services that are not currently outsourced are candidates for future growth [8]. It is estimated that approximately 70 % of RL services are performed by manufacturers. The market sector available for 3PL providers includes: product labeling and packaging; assembly; freight auditing and payment; transportation management; supply chain consultancy; customer service; lead logistics provider/fourth-party logistics provider services (LLP/4PL); order processing; and RL [8]. While 3PL providers typically offer these kinds of transportation management and consulting services, many of them are also beginning to offer RL services, because they recognize the market opportunity these services present [9]. In addition to their traditional services, some 3PLs now offer complete supply chain solutions for warehousing, order fulfillment, repackaging, re-labeling, assembly, light manufacturing, and repair [10].

A Closed-Loop Supply Chain (CLSC) system includes both forward and RL activities. RL is very different from the forward flow. However, RL entails processes similar to those employed in forward logistics such as information management of flows including tracking and credit of raw material, inventory, finished goods, and related information from the point of manufacture/distribution to the point of consumption or use. Finding and implementing effective RL within a CLSC system is a top concern of executives, managers, and other operational personnel. Managing return flows often require a specialized infrastructure to balance the relatively high handling costs and the additional processing time. As businesses try to control these costs, the demand for RL services from 3PLs increases [11]. 3PL providers specialize in logistical activities and are therefore able to offer to manufacturers, wholesalers, and retailers a number of attractive advantages. These advantages include but are not limited to transportation operations such as product return, which 3PL providers can perform at less expense than the original equipment manufacturers (OEMs). Manufacturers, wholesalers, and retailers outsource RL and repair functions to 3PL providers who, due to specialization and economies of scale, are often able to make greater investments in computerized logistics management systems and can configure their transportation systems to accommodate individual clients' needs [12]. Companies who purchase RL assistance from 3PL providers can reduce annual logistics costs by up to 10 % [13].

In principle, we can see the appeal of 3PLs for manufacturers. Yet, 3PLs face obstacles to penetration in the expanding market for RL services. One of these obstacles is confidence. 3PL providers who engage in RL operations can win the confidence of service receiving companies through high-quality service, risk and reward sharing, reduced distribution costs, service flexibility and responsiveness, managerial support, and understanding clients' supply chain needs [14]. Another way to increase confidence is the management and reduction of damages. Progress in these areas will lead manufacturers, wholesalers, and retailers to increase the use of 3PL providers for RL activities.

Moreover, it should be noted that it is difficult to repair a "damaged reputation". It is said that the complaint of a dissatisfied customer will affect negatively 30 people, while only three people will hear a satisfied customer's praise. The service quality and reputation of providers play significant role in 3PL selection by industry [15, 16].

The quality offered by a 3PL entails many distinct considerations such as on-time delivery, accuracy of order fulfillment, frequency and cost of loss and damage, promptness in attending to customers' complaints, and commitment to continuous improvement [17–19]. The market size is specific, and the pie grows very slowly. Yet, the number of those who want a piece of the pie increases very rapidly.

8.2.1 Damage and Reverse Logistics

One of the main activities of RL is product return management. This activity requires adequate quantities of used products of the right quality and price, and their availability at the right time. In terms of CLSC and RL, there are many different return types and volumes which occur during or even after the product life cycle, for which innovative methods of value recovery must be found. For instance, early in the product life cycle, commercial returns may be best used to fill warranty demands,

Table 8.1 Some of major appliance and consumer electronics product return rates[a] [12]

Category	Product Failure - defective (%)	Damaged in Shipping - defective (%)	Stock Balancing (%)	Shipper error - nondefective (%)	Customer return - nondefective (%)	Total[b]
Washers/ dryers	19	37	5	2	37	3.5
Televisions	32	12	18	16	38	18.6
Desktop Computers	35	10	14	12	28	23.6
Refrigerators	21	41	2	1	35	4.6

[a] Rates of manufacturer
[b] As a percentage of products shipped

Table 8.2 Consumer goods return experience (grade of damages) [12]

Category	Causes of returns	(Annual percentage) consumer electronics (brown) (%)	(Annual percentage) consumer appliances (white) (%)
Defective	Failure	9.4	1.5
	Shipment damage and other causes	2.3	1.2
Nondefective	Stock balancing and store returns	3.6	0.5
	Shipping errors	3.2	0.3
	Customer returns and dissatisfaction	7.5	1.0

whereas at the end of the life cycle, product returns may be best used to meet future demand for repair parts, after regular production has ceased. Determination of return types is an inherent element of RL, which also provides information on the return reasons. Damage is one of the return reasons in RL.

Damage caused to goods during the transportation process can have a negative effect on the relationship between 3PL providers and shippers. "Cases of damage" are instances in which a product is damaged in the hands of 3PL providers during the logistics process.

3PL providers are intermediaries who handle products which belong to a manufacturer or retailer rather than to themselves. Businesses depend on 3PL providers to ensure that products arrive in the condition they were in when they left the factory or retailer. Damage that occurs during the logistics process does not simply affect the reputation of a 3PL provider and its relationship with the shipper; in as much as a 3PL provider has direct access to the product itself damage also affects the reputation of the shipper and his relation to the customer. Besides damages, 3PL providers are concerned with other potential logistical problems including transportation delays, unfulfilled orders, repairs, and maintenance.

All products are at risk of sustaining damage during the logistics process, and damages sustained during the transportation process are one of the five most common reasons that goods enter the RL process. Table 8.1 shows the relative percentage of product returns for a number of products, along with the reasons for those returns. The "Damaged in Shipping" and "Shipping Error" columns represent the percentage of returns attributable to 3PL operations. In addition to consumer products, raw and semi-finished products may be also damaged during loading, transportation, and storage. Products can be damaged in both the forward and RL processes. Nevertheless, shipping errors for both defective and non-defective products account in total for 5.5 % of all types of damage annually [12].

While 3PL providers are responsible for the delivery of undamaged goods, they are also responsible for most of the damages that occur to goods during the logistics process. To control this cost, a growing number of 3PLs have begun to examine ways to improve the efficiency of product returns [20]. By reducing the number of cases of damage, 3PL providers can increase service quality and reduce the costs associated with damage in both forward and RL.

Table 8.2 shows the causes of returns in the United States in terms of defective and nondefective products. The total annual percentage of returns on consumer goods is 26 %.

Table 8.3 shows the market segments for which 3PL providers currently offer services, their activities, the type of 3PL providers, and the percentage of market that currently employs 3PL services. Technical products of high value include computers, industrial equipment, medical equipment, electronics, electrical equipment, and photo/printing products. Consumer products include food and sundry products, clothing/textiles, appliances/electronics, general merchandise/grocery, transportation/automotive, instruments, building equipment, and chemicals [12].

By providing high-quality service, 3PL providers who engage in RL operations can win the confidence of service receiving companies. This confidence will increase the exploitation of 3PL providers in RL. There have been many studies investigating success factors for 3PL partnerships, and these factors include [14]:

- service improvement
- reduced distribution costs
- service provider flexibility and responsiveness
- managerial support and understanding clients' supply chain needs
- risk and reward sharing.

An investigation of the causes of damage and the operations, in which damages are most likely to occur within 3PL providers' facilities, will enable the latter both to improve the quality of their service and to reduce costs to manufacturers, retailers, and customers. In order to understand the causes of damage and how damage may be prevented, this study uses information drawn from observations

Table 8.3 Key characteristics of major RL segments [12]

Product type	Reasons for return	Type of 3PL providers	Percentage of companies that use 3PL providers
Technical products of high value	Return of used or broken equipment and parts for repair, refurbishment, reclamation, or disposal	Asset-based logistics providers and logistics operations within repair depots	Manufacturer level: 33–35 % Wholesaling level: 24–25 %
Consumer products	Return of goods for disposal or disposition because of damage, expiration, seasonality, or similar reason f	Asset- or nonasset-based providers: the widest pool of 3PL types	Manufacturer level: 26–29 % Wholesaling level: 22–24 % Retailing level: 32–35 %
Green products	Validation of environmental regulatory compliance, recycling, and so on.	Niche specialists and waste specialists: not typical providers	40–42 %
Packing, pallets and containers	Management of parts of shipping containers and packing	Owners\mergers of container and pallet parts, trucking or rail companies	35–40 %
Waste disposal and remediation	Disposal of waste from commercial\industrial operations	Asset-based 3PLs; most specialize in waste disposal and related activities, rather than the hauling of goods	General: 70 % Specialized: 80–82 %

and investigations of the most frequently encountered cases of damage at 3PL providers.

8.3 Cases of Damage and Their Classification

This chapter proposes a new classification system through which cases of damage can be accurately and efficiently categorized to help 3PL providers better understand, manage and prevent cases of damage. As mentioned previously, damage that occurs during the logistics process is itself a reason that goods and products enter the RL process. As the proposed system will help manage and reduce damages, it will also reduce the need for damage-initiated RL processes.

Each case of damage may be classified using a three-part system, such as the one presented in Fig. 8.1. This system is based on the location where the damage occurs, the transportation and product type of 3PLs, and finally the contract and conditions. In other words, this system begins by establishing when the damage occurred, the type of product that was damaged, and the means by which damage is defined. Regardless of a 3PL provider's product specialization, all providers can use this system to classify their cases of damage. At the same time this classification may help the billing of damages.

The responsibility to protect products from damage belongs to the 3PL provider from the moment of the receipt of an intact product from the shipper. From that moment until its delivery, either the product remains on a pallet in the outbound loading dock, or in a truck on the road, or at an interim storage facility, or in an unloading area, the 3PL provider is responsible for ensuring that the product will not be damaged. The first part of our classification system is therefore to determine at which point in the logistics process a particular case of damage occurs in order to determine who is responsible for the damage. If the damage occurs at the manufacturer's warehouse, then the 3PL provider may not be liable. For example, before accepting the goods, a careful examination of the 3PL provider may result in a "received in damaged condition" liability exemption. The location in which damage takes place determines frequently whose insurance is liable for the damage. However, the aforementioned exemption applies only to damages specifically described, which means that the 3PL provider may yet be liable for any additional damages after the initial one, for which he is not responsible. Beyond such an exemption from liability, there are additional benefits in determining the point at which damage takes place. First, the point at which damage of forward-bound products occurs can determine the point at which they will enter the RL process, and how long they will remain there. Second, determining the locus of origin may alert 3PL providers to particularly problematic sectors of their logistics operations.

The second step in classifying damages is to determine the product type by its transportation type. The following is a list of the five product categories handled by Borusan Logistics.

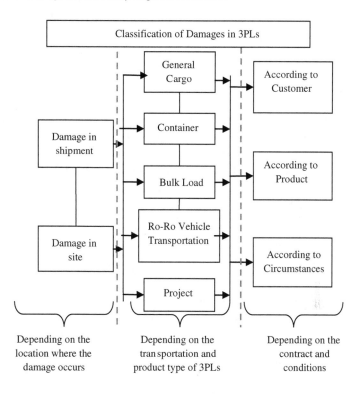

Fig. 8.1   Classification of damages in 3PLs

- General Cargo, including the carriage of cold/hot rolled sheet plates, packed sheet plates, and boxed products.
- Container, including 20' or 40' containers and open-top containers.
- Bulk load, which is usually utilized in shipments of large quantities of liquid chemicals (such as urea).
- Ro–Ro (roll-on/roll-off), which is a shipping category that includes vehicles and bulk items in wheeled containers.
- Project containers: the specific classification is given to specially produced products that require special vessels and processes for shipment.

Note that it is important to determine the product type, because different types of product are susceptible to different kinds of damages. For instance, a steel roll might suffer strike damage, ovalization, or broken welds, whereas containers might suffer tilt or drop damage. Moreover, the transportation methods employed will vary according to the type of product. Bulk load products may spill from tanker trucks, while project containers may be damaged by forklifts.

The third part of our classification system concerns the identification of damage by category. There are three primary ways of defining damages:

- Customer-defined damage: damage in this category may be difficult to anticipate. For example, automobile manufacturers may consider oil residue on rolled sheet steel to be "damage" while other manufacturers may not. Another example is the slight bending of rolled steel sheets. Whether or not the customer would consider this as "damage" depends on the customer's intended use of rolls.
- Product-defined damage: damage in this category is determined by the integrity of the product itself, especially packaged product. For example, a large hole or tear in the outer packaging of steel sheets may not be considered damage while similar damage to the packaging of a television set would be likely considered as damage.
- Circumstance-defined damage: typically, damage in this category is defined jointly by the shipper, the receiver, and often the 3PL provider. It may include products that must be delivered by a certain date or which are themselves time sensitive.

In the second and third categories, what constitutes damage is usually stipulated by a contract, which is meant to protect all involved parties. It is important, then, that the 3PL provider takes care to accurately record cases of damage, not only to improve operations, but to protect himself under the contract terms. The proposed damage classification system provides a more comprehensive way to record and understand damage.

8.3.1 Examples of Damage

In this section, a number of examples of recurrent damages that have occurred at Borusan Logistics are presented, so that our proposed classification is demonstrated. Moreover, we believe that they will be helpful for readers to understand what kind of damage may occur to items that are in the care of 3PL providers.

8.3.1.1 Cases of General Cargo Damage

The designation "general cargo" refers to the carriage of cold/hot rolled sheet plates, packed sheet plates, and any kind of boxed products. The following categories constitute instances of this kind of damage:

- Rust damage (Fig. 8.2): before the produced roll reaches the customer, it has undergone a further slicing procedure by a different firm. After this procedure, sometimes the roll rusts and becomes useless, because the wooden equipment put between the rolls during storage has caused rust.
- Damage as a result of falling out of a truck: this damage (Fig. 8.3) might result from an incomplete lashing of the roll during transport, sudden braking of the

Fig. 8.2 Rust damage on
the rolls

truck, difficulties in handling encountered by employees during loading and
unloading to the body of the carrier, or traffic accidents.

- Ovalization damage (Fig. 8.4): it may occur if the logistics firm (3PL) stacks
  items higher than the maximum value specified by the producer or customer.
- White rust damage: this kind of damage (Fig. 8.5) occurs when, for whatever
  reason (a) water enters the rolled sheet plate, and the material has been packed
  without allowing it to cool after production or (b) the hot/cold balance has not
  been established. Under these conditions, the material perspires, and the
  resulting water remains inside it, thereby turning into white rust.
- Tearing damage: this type of damage (Fig. 8.6) occurs mostly when the material
  gets caught on something, or when it touches the construction machine or a
  vehicle. Additionally, the damage may be caused by the equipment used to lash
  the material during its transfer or the journey. In this case, the material should be
  accepted for production after the damaged part is separated.
- Winding fluctuation damage: such damage occurs mostly in the holds of ships
  either under adverse weather conditions during the journey or when lashing is
  not secure (Fig. 8.7). Usually many rolls are damaged at the same time, causing
  substantial losses. Entire shipments can be damaged this way.
- Scratch damage: these cases of damage may result from a construction machine;
  especially a forklift. The damaged part must be separated from the roll, so that
  the remainder can be utilized in parts of secondary quality.
- Damage to flashing: the equipment hitting the roll does not deeply damage the
  winding; however, it causes a flash-like effect on the lateral surfaces of the
  winding shown in Fig. 8.8, because it passes by rubbing. Unless the firm doing
  the production works with zero tolerance, the damage might be compensated.
  In such cases, there is generally a high possibility to use the roll without sep-
  arating the damaged part.

Fig. 8.3 Falling out of a
truck

Fig. 8.4 Ovalization
damage

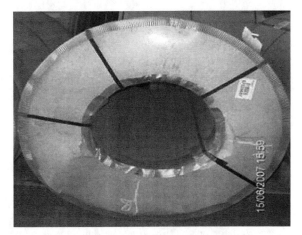

Fig. 8.5 White rust damage

Fig. 8.6   Tearing damage

Fig. 8.7   Damage of fluctuation of winding

Fig. 8.8   Damage of flashing

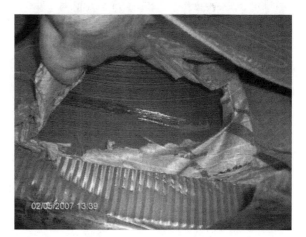

Fig. 8.9 Damage of being
hit

- Salt water damage: such cases of damage occur if water enters the warehouses, when vessels suffer from adverse weather conditions during the journey. They are revealed using the Silver Nitrate test.
- Damage to external packaging: this damage usually occurs during handling. It is highly probable to see this damage while placing the roll on the vessel, during its unloading from the vessel or if it is transferred inside the depot.
- Damage resulting from the displacement or disconnection of the hoop: damage of disconnection of the hoop does not generally inflict much damage on the rolled material. However, the package will open if more than one hoops disconnect.
- Damage caused by tilting on the body: this causes the packed sheet plates to tilt in cases of sudden braking or shaking.
- Crush damage: in this type of damage the packed sheet plate might be crushed or scratched due to the equipment or machine that has hit the packed sheet plate. Upon the separation of the damaged leaves, the remaining parts can be used.
- Bending damage: during handling, the pipes in bundles get caught on a sling or another apparatus. As a result, they are bent and become scrap.
- Crushed pipe mouthpieces: these cases of damage might occur mostly as a result of falling or by being crushed by something heavy as seen in Fig. 8.9. The pipe then becomes useless, because its joint is damaged.
- Scratched and rusted pipes: these cases of damage are generally caused by the use of wrong equipment during loading and unloading. Initially, scratches occur on the surface of the pipe and then these scratches rust. Thus, the pipes become useless.

8.3.1.2 Cases of Container Damage

This section deals with cases of damage to containers, the most moved items at the ports.

- Strike damage: this damage (Fig. 8.9) generally occurs when the content of a container gets free from its fastenings, moves, and strikes the walls of the container.
- Tear damage: this type of damage occurs when the container is hit by equipment, such as cranes, forklifts and wreckers. Repair is affected by cutting the sheet plates at the tear site with an oxygen welder and welding new plates to that site to complete the repair process.
- Burst damage: the material in the container puts pressure on the lateral walls and causes the walls to project outwards. This differs from impact damage in that the main posts of the container are not harmed. In the case of impact damage, however, the container is exposed to torsion.
- Convex damage: this damage generally occurs when a construction machine hits the container or when a stationary container is hit during the stuffing process.
- Concave damage: as opposed to convex damage, in this kind of damage, the walls of the container are bent inwards as a result of an exogenous blow to the container.
- Hole damage: this damage is caused by a machine or a part of a machine. The damage is rectified by replacing the damaged part.
- Small hole damage: it is classified differently from hole damage because of its smaller size. The important thing is that these holes cannot be realized owing to their small size.
- Oblique damage: this type of damage is generally seen on the door of the container. It often arises from the bending of the door bars. The damage is rectified by the replacement of the damaged parts.
- Scratch damage: this kind of damage receives its name from the fact that it is characterized by a scratch that originates near the right side of the goods and becomes deeper as it progresses. It generally results from friction or rubbing that occurs during contact with a machine or other structure while passing without appropriate clearance. To correct this type of damage, the damaged parts are disassembled and repaired or replaced.
- Breakage: generally, breakage occurs when the locking and placing parts of a container are broken. The damage is corrected by replacing the broken part or through repair by welding.
- Broken Seals: under Customs law, incoming containers must remain sealed until delivery. Additionally, the numbers of these seals are specific. Records should be kept of incoming containers without seals, applicable procedures undertaken, and customs notified.
- Loss: this kind of damage occurs when an apparatus on the container is displaced and falling over. The damage is rectified by the replacement of the lost part with a new one.

Fig. 8.10 Project damage

- Refrigeration leakage: these cases of damage occur on refrigerated, or "reefer" containers. Damage may prevent the product inside from remaining cold. If this happens, spoilage may occur.

8.3.1.3 Damage in Project Loads

These cases of damage occur in specific project loads. Parts are produced specifically for a project and should they be damaged the liabilities for the damage are great, and the consignee will be late or unable to complete its operation. In this case (Fig. 8.10), the parts should be repaired immediately or new parts should be produced and sent to customer.

8.3.1.4 Damage to Vehicles/Construction Machines/Equipment

This section examines instances of damage to the indispensable logistics vehicles, including cranes, ships, TIR trucks, and forklifts. The damage to construction machines is the most serious kind of damage that might happen, not only because of the loss of equipment, but because of the loss of products, structures, and even the human injury that may be involved. Further, logistical activities must cease until damage is assessed and corrected, and replacement machines are brought in. Apart from these visible costs, logistical enterprises face many invisible costs such as the time lost to cessation, waiting, and restarting, as well as worker slowdown and substitution.

- Damage to construction machines-1: this category of damage includes accidents that happen to a vehicle called a "Stacker", which is used to handle containers

Fig. 8.11 Damage to
construction machine

at ports. As seen in Fig. 8.11, both the vehicle and the container are damaged. This kind of accident often occurs while lifting unbalanced loads.

- Damage to international road transport (TIR) Trucks: this damage takes place on highways or within facilities. Not only is the vehicle itself damaged, but often its load can be damaged. Such cases of damage are generally resolved through the involvement of several types of insurance, including carrier's liability insurance, transportation insurance, and so on.
- Damage to construction machines-2: at ports, a machine called rubber tyred gantry (RTG) is used to stuff containers and load them onto vehicles. The damage presented in Fig. 8.12 occurred when the wrecker positioned below the RTG moved before the RTG unlocked its locks during the loading of a vehicle.
- Damage to Construction Machines-3: this type of damage is caused by the imbalanced loading of containers during stuffing. The resulting imbalance led to the container's tilting and collapse (Fig. 8.13.).

8.3.1.5 Damage to Boxed Products and Bulk Loads

This type of damage is caused by construction machines or other equipment. If the boxed product is a single piece (i.e., oven or refrigerator), it becomes useless. However, if the product consists of separate products, as in the example provided in Fig. 8.14, damaged units are separated out and the rest ones may be used. If the product spills, it is important to identify its chemical or physical properties before any intervention takes place.

Fig. 8.12 Damage of construction machine-2

Fig. 8.13 Damage of construction machine-3

Fig. 8.14 Damage of a bulk load

8.4 Damage Evaluation Procedures: An Overview

The general procedures proposed in this section should be followed in order to properly evaluate any case of damage. However, depending on the situation in which damage has occurred, certain other procedures may be called for. The evaluation of a case of damage involves three processes:

- The examination of the company's record as well as the documentation of the damage,
- Filling an insurance claim, and
- The undertaking of preventative actions to avoid further damage.

As damage may occur during the regular operations of a 3PL provider—including loading, unloading, and transfer—damage evaluation procedures are, or at least should be, a fundamental part of their operations model.

The most important consideration of a 3PL should be the prevention of damage. Consequently, it is crucial to identify the points at which damage might occur during the logistics process. To properly plan such procedures, damage evaluation is very important; the accurate documentation and description of a damage that has occurred will help managers to identify, classify, and construct a systemic understanding of patterns of damage in order to better identify their likelihood and avoid their future realization.

The insurance process is initiated when the appropriate departments of the 3PL provider are notified about the damage that has occurred. If a 3PL fills an insurance claim, the insurance company assigns a risk expert to assess the damage. Whether or not to obtain information from the insurance company depends upon this person's decision regarding exemption and liability. The aim of a 3PL should be to report a damage as quickly and efficiently as possible. Reports on cases of damage must provide the Department of Financial Affairs/Insurance (Department of Insurance) with all relevant details. These details ensure that all the necessary rescue and protection measures will be taken. Such measures should be performed no matter if the damage is covered by insurance, in order for a 3PL to avoid injury of staff or third parties, and order to protect its assets.

The insurer requires always proof and documentation of the damage claim. Therefore, except for specific compulsory or emergent cases, no changes should be made without the insurance company's approval either to the site at which the damage occurred or to the goods themselves. It is the 3PL company's responsibility to provide the necessary documents. The latter includes official governmental documents such as an official fire department report in the case of fire damage, a police report in the case of theft or an inspector's report prepared by Social Security Institution inspectors in the case of a work accident, as well as company accident reports and signed witness statements. Other necessary information provided by these documents are the time and date, the type of damage (i.e., fire, storm, water, etc.), the cause of damage, the equipment involved, the extent of damage, and contact information for the damage expert and other

Fig. 8.15 Proposed
organizational chart for
managing damage

Fig. 8.16 Distribution of
damage by region between
January and December 2006

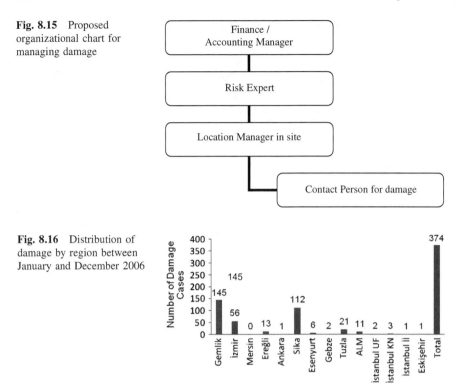

appropriate agencies. Because of their importance to the evaluation process, it is
crucial that such reports are made after any accident occurs. The risk expert's job
is to inspect the damage and submit a prepared "expert report" to the insurer. A
prompt evaluation of a damage should be able to answer questions regarding its
cause and should ensure that information will be ready before the insurance
company asks for it. Otherwise, the process may take more time. All documents
must be sent to the 3PL provider's insurance department, though not all of them
may be needed by the insurance company. Moreover, depending on the conditions
of insurance, no oral or written statements expressing responsibility for damage
should be made without the prior approval of the insurance company.

The implementation of information technology has a great affect upon damage
prevention and the evaluation procedures presented previously. Basic logistics
information systems are composed of order management systems (OMS), warehouse management systems (WMS), and transport management systems (TMS)
[21]. Logistics operators must keep up with the always changing technology. For
example, digital photography has substantially shortened the amount of time it
takes to examine and evaluate cases of damage. Electronic filling and e-mail has
significantly increased the speed and efficiency of documentation procedures.
Businesses no longer need to rely on individuals making arbitrary classifications

and decisions. These arbitrary decisions constitute more complex and unmanageable processes for companies. Although a company's relief from dependence on individuals is usually replaced by a dependence on information technology, the benefits of an electronic filling system—faster, more accurate, accessible, uniform, and adaptable data processing—far outweigh the relevant costs.

8.4.1 Implementing a Damage Classification at Borusan Logistics

From 2005 to 2007, the proposed damage classification system was used to evaluate and analyze cases of damage at Borusan Logistics in Turkey. It was used by the risk expert who worked on behalf of the company as the decision maker and controller of all incidents of damage (Fig. 8.15), keeping detailed records of cases by location, month, type of product, and cause of damage in this new damage control organizational system through risk assessment, and management structure.

Before this study, Borusan Logistics was only concerned with controlling large-scale cases of damage and the cases which were expressly covered by insurance. Damage was not analyzed systematically, and there were often incidents of missing documentation, problems in monitoring the damage process, few efforts toward improving the process, and no company investigations regarding accidents. Through the implementation of the suggested procedure and organizational structure, Borusan Logistics managed to investigate and control damages, while it has become more proactive toward developing both solutions for damages as well as methods of preventing them.

Before the implementation of the above process, the primary document for recording damage was an accounting report which described the nature of the damage, the responsible party, as well as the damage cost. The insurance company would designate its own risk expert to assess damage when the company filled a claim. On the one hand, this organizational structure had serious flaws: for example, after a damage incident, the company needed to wait for the insurance company's risk expert before taking remedial action, and such delays could be especially costly in emergency situations. On the other hand, the concentration of information and management on a single position enabled the coordination of departments and streamlined processing. Most importantly for the purposes of this study, our documentation of cases of damage was facilitated by the fact that Borusan Logistics' operations departments were familiar with a centralized reporting structure. Operating as Borusan Logistics' risk expert, we were able to collect data on cases of damage for the construction of the charts presented subsequently.

The distribution of cases of damage by region is provided in Fig. 8.16. The chart shows the amount of damage occurring at each location. Some locations clearly exhibit higher damage rates than others. Figure 8.17 shows the number of cases of damage that occurred throughout the company in 2006. There is clearly a seasonal trend.

Fig. 8.17 Distribution of
damage by month between
January and December 2006

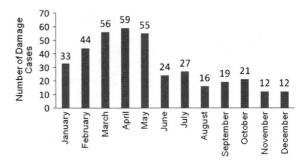

Fig. 8.18 Distribution of
damaged products between
January and December 2006

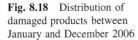

Fig. 8.19 Causes of
damages between January
and December 2006

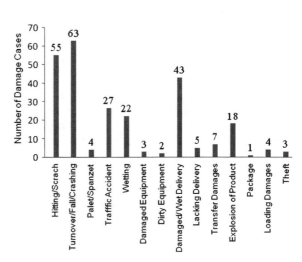

Figure 8.18 charts the types of products that were damaged in 2006. It can be clearly seen which products are most likely to sustain damage. Finally, Fig. 8.19 charts the causes of damage, indicating that certain causes are more common than others.

Using the above data, it was possible to draw useful correlations in order to determine the systemic causes of damage and potential preventative solutions. For example, a correlation between the seasonal variation of damage and a specific location suggests that open air facilities had a higher risk of damage incidents. If this connection is further correlated with high rates of water damage at that location, then this further strengthens this conclusion. A possible solution is to construct more covered warehouses.

Other correlations include those established between a specific location and the mode of transportation used or between the type of product, e.g. milk, and the type of damage, e.g. spoilage, which might suggest better refrigeration. Through such procedures, patterns of damage and their possible prevention become apparent.

Other variables which contribute to cases of damage, such as new specifications of customer-defined damage, may require the addition of further classification categories. Also, the location of damage category might be divided more specifically by department. If costs associated with damage are associated with the damage categories, department managers can perform cost-benefit analyses when considering damage control procedures.

8.4.2 Suggestions and General Solution Proposals

In order for a 3PL to employ this classification system, it will be necessary to ensure that all damage cases are promptly reported. This requires that both companies and their employees change their attitudes toward damage, which should no longer be considered as a normal or inevitable business cost. A 3PL may need to redefine certain job positions in relation to the damage documentation process and provide to the people in these positions the appropriate training regarding both damage documentation and insurance processes. In some cases, ongoing training programs should be implemented. For example, a weekly training at a depot on how to report damage cases—for example, when products fall from forklifts—will ensure that employees are knowledgeable of damage reporting procedures. To encourage the proper and prompt reporting of damage, punishment should not be used as a motivational tool. Rather, a reward system for damage-free operations is preferable. In addition, an efficient, up-to-date, integrated computer infrastructure should be constructed for efficient and accurate tracking of damages. 3PL providers might also designate a group of employees and transportation systems to provide customized services to a particular company. This way, employees will become experts at that company's requirements, which should result in further reductions of cases of damage as well as cost savings for the customer, the 3PL provider and the insurance company.

Perhaps the most important aspect of the proposed damage evaluation procedures is their use by a company-designated risk expert in conjunction with the damage classification system. Instead of reporting to the accounting office, damage-prone departments will need to report to the designated risk expert. Each depot should have a designated risk expert who will be in charge of evaluating and recording all cases of damage. Sustainable communication among these experts is the effective way to share information from depot to depot. A 3PL provider's risk expert position will make damage evaluation more accurate and efficient, expedite the filling of insurance claims, and refine the stipulations of liability contracts. Especially, the latter can save manufacturers, 3PL providers, and customers the expense associated with insurance claims and other liability costs.

8.5 Conclusion

Manufacturers can gain an advantage over the competition by availing themselves of the increased customer service and delivery quality, as well as the lower transportation and production costs offered by 3PL providers in RL and CLSC systems. However, damages which occur while goods are in the possession of 3PL providers can undermine the recognition of 3PL providers as cost-saving and service improving enterprises. Therefore, it is up to the 3PL providers to build confidence in the 3PL industry by reducing incidents of damage while goods are in their possession. To this end, the authors propose a system of classifying and categorizing cases of damage in order to discover patterns and correlations in the incidents of damage which can then be used to determine appropriate systematic damage prevention solutions. The implementation of this classification system can be most effectively implemented through a 3PL provider's designation of its own risk expert. By employing the proposed classification system in conjunction with a designated risk expert, a 3PL provider can reduce cases of damage during both the forward logistics and RL processes. This model may be adapted as necessary to the needs of any 3PL provider, whether in the transportation of raw or finished products. Finally, the model is meant to provide a new theoretical approach to the evaluation of cases of damage.

Acknowledgments This work was conducted as part of the MBA Project based at the School of Business of the University of Sakarya. The authors would like to thank Borusan Logistics Company for research support. We would like to thank the editor for his constructive comments on our chapter. We also would like to thank Brent Curdy for his interest and thoughtful comments regarding this chapter.

References

1. http://www.rlec.org/
2. Dowlatshahi S (2000) Developing a theory of reverse logistics. Interfaces 30(3):143–155
3. Fleischmann M, Krikke HR, Dekker R, Flapper SDP (2000) Characterization of logistics networks for product recovery. Omega 28:653–666

4. Guide VDR Jr, Van Wassenhove Luk N (2009) The evolution of closed-loop supply chain research. Oper Res 57(1):10–18
5. Brito MP, Dekker R (2002) Reverse logistics—a framework. Econometric Institute Report EI 2002-38
6. Hertz S, Alfredsson M (2003) Strategic development of third-party logistics providers. Ind Mark Manag 32(2):139–149
7. Bloomberg DJ, LeMay S, Hanna JB (2002) Logistics. Prentice-Hall, Upper Saddle River
8. Results and Finding of the 14th Annual Study (2009) The state of logistics outsourcing in 2009. John Langley, Jr., Ph.D., and Capgemini U.S. LLC
9. Prahinski C, Kocabasoglu C (2006) Empirical research opportunities in reverse supply chains. Omega 34(6):519–532
10. Wang Y, Sang D (2005) Multi-agent framework for third party logistics in E-commerce. Expert Syst Appl 29(2):431–436
11. Green FB, Turner W, Roberts S, Nagendra A, Wininger E (2008) A practitioner's perspective on the role of a third-party logistics provider. J Bus Econ Res 6(6):9–14
12. Blumberg DF (2005) Introduction to management of reverse logistics and closed -loop supply chain processes. CRC Press, Boca Raton, pp. 23–43
13. Krumwiedea DW, Sheu C (2002) A model for reverse logistics entry by third-party providers. Omega 30(5):325–333
14. Selviaridis K, Spring M (2007) Third party logistics: a literature review and research agenda. Int J Logist Manag 18(1):125–150
15. Lynch CF (2000) Managing the outsourcing relationship. Supply Chain Manag Rev 4(4):90–96
16. Boyson S, Corsi T, Dresner M, Rabinovich E (1999) Managing effective third party logistics relationships: what does it take? J Bus Logist 20(1):73–100
17. Razzaque MA, Sheng CC (1998) Outsourcing of logistics functions: a literature survey. Int J Phys Distrib Logist Manag 28(2):89–107
18. Langley CJ, Newton BF, Tyndall GR (1999) Has the future of third party logistics already arrived?. Supply Chain Manag Rev, Fall, pp 85–94
19. Stock GN, Greis NP, Kasarda JD (1998) Logistics strategy and structure—a conceptual framework. Int J Oper Prod Manag 18(1):37–52
20. Mina H, Kob H (2008) The dynamic design of a reverse logistics network from the perspective of third-party logistics service providers. Int J Prod Econ 113(1):176–192
21. Kim C, HoonYang K, Kim J (2008) A strategy for third-party logistics systems: a case analysis using the blue ocean strategy. Omega 36(4):522–534

Index

B
Basel convention, 79, 80
Business plan, 47, 48

C
CEN Workshop Agreements, 41, 47, 48
CEN-European Committee for Standardisa-
    tion, 40, 46
Closed loop supply chain, 21, 31, 55
Consumer electronics, 26, 73, 81, 135
Control chart(s), 2, 29, 120, 123, 125, 126
Corporate social responsibility, 47, 53, 54, 69
Corporate sustainability, 61

D
Damage
    classification, 140, 151, 153
    definition, 132–136, 140, 141, 144, 147,
        153
    evaluation, 149, 153, 154
    examples, 28, 140
Disassembly, 9, 10, 12, 15, 23, 29, 32, 33, 43,
    57, 60, 74, 46, 81, 92, 96–104,
    108–114
Disposal, 3, 7, 10, 42, 59, 74, 78, 79, 81, 87,
    88, 90, 96, 119, 133

E
End of life management, 74
End-of-life product, 56
Environmental management, 30, 43, 47,
    55–57, 61, 63, 65, 79, 81–83, 91
E-waste, 75, 76, 78–81, 83,
    87, 88

F
Field study, 75, 90

G
Global reporting initiative, 55
Grading, 12, 23, 30

I
Indicator(s)
    economic, 25, 54, 57, 58, 60–65, 67, 69,
        70, 76, 79, 89, 99, 132
    environmental, 15, 56, 63
    performance, 55, 56, 62
    social, 55, 62, 63, 65–67
Information technology, 7, 8, 10, 16, 48, 73,
    116, 150, 151
Inventory-ies, 3, 4, 6, 8, 9, 13, 14, 44, 46, 59,
    74, 96, 98, 110, 114
ISO 14001, 82, 86
ISO 9001
    ŋ ISO 9000, 2, 47, 81, 82, 86

L
Legislation, 5, 6, 45, 47–49, 54, 83, 86, 88, 89
Linear programming, 8, 10, 57

M
Mixed integer programming, 10, 60, 98

O
Original equipment manufacturer(s), 3, 30, 86,
    134

Y. Nikolaidis (ed.), *Quality Management in Reverse Logistics*,   157
DOI: 10.1007/978-1-4471-4537-0, © Springer-Verlag London 2013

P

Product
  acquisition, 8, 9, 12
  design, 2, 8, 14, 15, 23, 98
  recovery, 3, 7, 8, 10, 11, 13, 14, 22, 24, 27,
    29–32, 58, 96–100, 109, 114, 116, 127
Product embedded information device, 99
Product recovery types
  cannibalisation, 4, 29, 43
  recycling, 13, 14
  refurbishing, 4, 29
  remanufacturing, 3, 4, 10, 12–14, 23, 24,
    27–29, 32, 35, 98, 100, 102
Production planning, 8, 10, 13, 14, 60

Q

Quality
  definition, 22, 27, 31, 34
  dimensions, 22, 25, 27, 35
  framework, 22, 32, 33, 35, 67
  manufacturing based, 25, 34
  product based, 24–26, 28, 30, 31
  uncertainty, 9, 23, 24, 26, 28, 33, 35, 36
  user based, 25, 30, 34
Quality assurance, 46, 74–76, 78, 79, 81–87,
  89–91, 96, 98
Quality control, 29, 30, 46, 116, 117, 123, 126,
  127, 132
Quality management, 22, 25, 30, 35, 62, 74,
  81, 82, 84, 86, 89, 91, 96, 115, 117,
  120, 127, 132

R

Radio-frequency identification-RFID, 11, 16,
  22, 97, 99, 100, 105, 109, 115–117, 120
Recycling systems, 74, 75, 83, 87, 90
Remanufacturing-to-order, 12–14, 24, 28, 29,
  32, 95, 97, 100
Reverse logistics
  environmental, 43, 45, 49, 54–56, 69, 84,
    85, 89

S

Sampling procedure(s), 46
Sensor embedded products, 96, 99
Sorting, 12, 14, 76, 78, 96, 133
Standard(s), 2, 9, 28–30, 40, 41, 44, 61, 76, 81,
  90
Standardisation, 40–42, 45, 47, 49, 50, 116
Supply chain
  management, 3, 21–23, 26, 42, 43, 56–58,
    75, 82, 86, 88, 116, 132
Sustainable development, 53, 62

T

Testing method(s), 39, 40, 46, 47, 49
Third-party logistics, 44, 48, 61, 86, 89, 132,
  133

V

Value creation, 21, 22, 33, 43, 74

Printed by Publishers' Graphics LLC
BT20130108.19.20.25